A FACTOR MODEL
APPROACH TO
DERIVATIVE PRICING

A FACTOR MODEL APPROACH TO DERIVATIVE PRICING

James A. Primbs

California State University
Fullerton, USA

CRC Press
Taylor & Francis Group
Boca Raton   London   New York

CRC Press is an imprint of the
Taylor & Francis Group, an **informa** business

A CHAPMAN & HALL BOOK

CRC Press
Taylor & Francis Group
6000 Broken Sound Parkway NW, Suite 300
Boca Raton, FL 33487-2742

Printed on acid-free paper
Version Date: 20161013

International Standard Book Number-13: 978-1-4987-6332-5 (Hardback)

Visit the Taylor & Francis Web site at
http://www.taylorandfrancis.com

and the CRC Press Web site at
http://www.crcpress.com

To my family,
Keri, Michael, Therese, and Patrick

Contents

List of Figures

Preface

The purpose of this book is to present the conceptual underpinnings of derivative pricing from a single, unified perspective—the factor model perspective. The key idea is that the stochastic differential equation models used in derivative pricing can be interpreted as linear factor models that relate sources of risk (the factors) to returns. When viewed in this way, an application of Ross' Arbitrage Pricing Theory (APT) leads directly to the absence of arbitrage equations that govern the pricing of derivatives. In my opinion, this is the simplest route to exposing the majority of the equations that arise in derivative pricing to someone relatively new to the field.

This book focuses on establishing that factor model framework and how it can be applied, in a rather generic sense, to derivatives. While the pricing of many derivative securities is covered from this point of view, the content is certainly not encyclopedic, and no attempt is made to touch upon every single contract in the market. Rather, this book is deliberately intended to have a long time constant. By that, I mean that while the popular derivatives of the day may change, the principles that underlie the pricing of those derivatives will not. This book is oriented toward explaining those rather timeless principles from the factor model perspective.

As a course, this book is aimed toward an introductory graduate or advanced undergraduate student who already has an exposure to the basic mechanics of options, futures, forwards, and the other derivatives that are usually covered in a first course. For example, the reader is assumed to be familiar with basic option terminology, static arbitrage relationships such as forward-spot price relationships, interest rate quoting conventions, etc. The appeal of using this book as a text for a second course is that it allows for coverage of many different derivatives models, but without relying on extensive knowledge of stochastic calculus, martingale theory, or other daunting prerequisites that are often emphasized before explanations of such models are broached. In fact, an advanced undergraduate student in engineering or the sciences with background in calculus, linear algebra, differential equations, and basic probability or stochastic processes, should be able to read and benefit from this book. Additionally, researchers and advanced graduated students who desire an understanding of the simple underpinnings that hold together the vast array of derivative pricing models—past, present, and future—will also be well rewarded.

In general, the book is written in an informal conversational style that emphasizes financial concepts over mathematical rigor. After all, derivative pricing is an applied subject, and at the end of the day I want the reader to not lose sight of that through the often dense forest of mathematical details that can clutter one's view. Moreover, this book seeks to unify thinking about derivatives under a single coherent theme, leading to a rather tight presentation. This is a deliberate choice. In any book, there is a trade-off to be made between breadth of coverage and focus. In this book I have chosen the side of focus, but in the hopes that depth of understanding of the factor model approach will lead to a kind of breadth of understanding of the foundations of derivatives all its own.

Finally, I would like to thank all of those who contributed to the vision of this book. Over the years, every interaction and piece of feedback I have received has led to improvements. Specifically, I would like to thank Sunil Nair, Alex Edwards, and Robin Lloyd-Starkes from CRC Press and Taylor & Francis Group. And last, I would like to thank my family—Keri, Michael, Therese, and Patrick—for allowing me to "hide away" while this book was being prepared.

James A. Primbs
May 2016

Notation

SYMBOL DESCRIPTION

$\mathbb{R}$ real line $(-\infty, \infty)$.

$\mathbb{P}(A)$ probability of set A.

$\mathbb{E}[X]$ expectation of X.

$\mathbb{E}^*[X]$ risk neutral expectation.

p^* risk neutral up probability

$\mathcal{N}(\mu, \sigma^2)$ Gaussian (normal) random variable with mean μ and variance σ^2.

$S(t)$ stock price at time t.

c_t $\frac{\partial c}{\partial t}$, partial derivative of c with respect to t.

c_x $\frac{\partial c}{\partial x}$, partial derivative of c with respect to x.

c_{xx} $\frac{\partial^2 c}{\partial x^2}$, second partial derivative of c with respect to x.

$z(t)$ Brownian motion.

$\pi(t; \alpha)$ Poisson process with intensity α.

x^T transpose of vector x.

A^T transpose of matrix A.

$x(t^-)$ limit from the left: $x(t^-) = \lim_{h \uparrow t} x(h)$.

x^- notation for limit from the left: $x(t-)$.

$\mathcal{P}$ vector of prices.

$d\mathcal{V}$ per unit value changes.

r vector of returns.

r_0 constant risk-free rate of interest (when allowed to be a function of time see short rate process below).

$r_0(t)$ instantaneous short rate process at time t.

λ_0 market price of time.

λ market price of risk.

$f(t|T)$ time t futures price for contract with delivery at time T *or* instantaneous forward rate at time t between times $T - ds$ and T.

$F(t|T)$ time t forward price for contract with delivery at time T.

$B_0(t)$ time t value of money market account.

$B(t|T)$ time t price of a zero-coupon bond that pays \$1 at maturity T.

$D(t|T_1, T_2)$ time t discount factor for value at time T_1 of \$1 paid at T_2.

$r(t|T)$ spot rate to time T at current time t under continuous compounding.

$R(t|T)$ spot rate to time T at current time t under simple interest.

$f(t|T_1, T_2)$ time t forward interest rate between time T_1 and T_2 under continuous compounding.

$F(t|T_1, T_2)$ time t forward interest rate between time T_1 and T_2 under simple interest.

CHAPTER 1

Building Blocks and Stochastic Differential Equation Models

T HIS chapter contains an introduction to the basic mathematics required for derivative pricing and financial engineering. Our approach is to construct models of asset prices from basic building blocks. In particular, we begin with the two building blocks of Brownian motion and Poisson processes. We then create more complicated models by using Brownian motion and Poisson processes to drive differential equations, which are then known as *stochastic differential equations*. Stochastic differential equation models play a central role in the factor approach.

Additionally, because Brownian motion and Poisson processes form the basis for our models, it is important to have a good feel for their essential characteristics. For this reason, we also provide simple binary approximations that highlight their key features and differences.

Some background knowledge in probability and stochastic processes is assumed. Moreover, the presentation here is mainly tutorial and heuristic. However, don't let that fool you. Having the proper intuition into the basic building blocks used to assemble stochastic differential equations will go a long way toward deepening your understanding of the models used in derivative pricing.

1.1   BROWNIAN MOTION AND POISSON PROCESSES

Brownian motion and Poisson processes are our fundamental building blocks for creating models of asset prices. The key features are that Brownian motion has continuous sample paths (with probability 1), and Poisson processes jump! We begin with Brownian motion which is built on the Gaussian random variable.

1.1.1 Gaussian Random Variables

A one-dimensional Gaussian random variable (also called a *normal* random variable) is a random variable X with density function,

$$X \sim f_X(x) = \frac{1}{\sqrt{(2\pi\sigma^2)}} \exp\left(-\frac{(x-\mu)^2}{2\sigma^2}\right), \tag{1.1}$$

where the notation $X \sim f_X(x)$ is used to indicate that X is distributed with density $f_X(x)$. In this case, $\mu \in \mathbb{R}$ is the mean of X and $Var(X) = \sigma^2 \in \mathbb{R}$ is the variance of X, defined as

$$\mu = \mathbb{E}[X], \quad \sigma^2 = \mathbb{E}[(X-\mu)^2], \tag{1.2}$$

where $\mathbb{E}[\cdot]$ denotes expectation. The standard deviation, represented by σ, is defined as the square root of the variance.

The Gaussian random variable plays a key role in probability theory due to the Central Limit Theorem which, roughly speaking, says that properly scaled "sums of random variables look Gaussian." In finance, the price of a stock, for instance, can be thought of as the sum of many random occurrences, such as the arrival of new information about a company and various buy and sell orders. Thus, it is not surprising that Gaussian distributions underlie many of the standard models of asset price movement.

To model more than a single stock, we will need the notion of an n-dimensional Gaussian vector. That is, an n-dimensional Gaussian (normal) random variable is a random vector $X \in \mathbb{R}^n$ with density function

$$X \sim f_X(x) = \frac{1}{(2\pi)^{\frac{n}{2}}|\Sigma|^{\frac{1}{2}}} \exp\left(-\frac{1}{2}(x-\mu)^T\Sigma^{-1}(x-\mu)\right), \tag{1.3}$$

where $(\cdot)^T$ signifies matrix/vector transpose, $\mu \in \mathbb{R}^n$ is the mean vector and $\Sigma \in \mathbb{R}^{n \times n}$ is the covariance matrix defined as

$$\mu = \mathbb{E}[X], \quad \Sigma = \mathbb{E}[(X-\mu)(X-\mu)^T]. \tag{1.4}$$

The term $|\Sigma|$ denotes the determinant of the covariance matrix.

We will use the notation $X \sim \mathcal{N}(\mu, \Sigma)$ to signify a Gaussian random variable or vector with mean vector μ and covariance matrix Σ. For example, using this notation, the standard Gaussian random variable (mean zero and variance one) is indicated by $X \sim \mathcal{N}(0, 1)$. More details regarding the Gaussian random variable can be found in standard probability textbooks such as [15] or [19].

1.1.2 Brownian Motion

Brownian motion (also known as a Wiener process) is a stochastic process built upon the Gaussian random variable. The idea is that each increment of the process in time is distributed as a Gaussian random variable with mean 0 and variance proportional to the time increment. Specifically, a real-valued stochastic process $z(t): t \in \mathbb{R}^+$ is a Brownian motion if:

1. $z(0) = 0$.

2. $z(t) - z(s) \sim \mathcal{N}(0, t - s)$ for $t > s$.

3. $z(t_2) - z(t_1), z(t_3) - z(t_2), \ldots, z(t_n) - z(t_{n-1})$ are independent for $t_1 \leq t_2 \leq \cdots \leq t_n$.

The first defining property states that a Brownian motion always starts at zero. The second property indicates that its increments have mean zero and a variance that grows proportional to the time difference, while the final property is that non-overlapping increments are independent.

You should remember the following facts about Brownian motion, as they make Brownian motion an ideal building block for unpredictable but *continuous* asset price movements:

- There exists a version of Brownian motion that has continuous sample paths.

- Brownian motion is nowhere differentiable with probability 1.

The first bullet above says that Brownian motion is appropriate for price processes that don't jump. In many cases, price processes do jump; hence we will need to introduce the Poisson process next to model jumps.

The second bullet can be interpreted in the context of predictability. If a curve is differentiable at a point it can be approximated locally by a line, with the slope of the line being the derivative of the curve at that point. But this means that we can predict (to order dt) the future value of the curve. In finance, we often want to assume that we cannot predict future prices. Non-differentiability indicates that in the sense mentioned above, future prices are not predictable.

Therefore, Brownian motion is an ideal building block upon which to create continuous asset price processes. A typical sample path of Brownian motion is given in Figure 1.1.

Just as there are vector Gaussian random variables, we can define a vector Brownian motion as follows. A vector Brownian motion $z(t) \in \mathbb{R}^n$ with covariance structure $\Sigma \in \mathbb{R}^{n \times n}$ is a stochastic process satisfying:

1. $z(0) = 0$.

2. $z(t) - z(s) \sim \mathcal{N}(0, \Sigma(t - s))$ for $t > s$.

3. $z(t_2) - z(t_1), z(t_3) - z(t_2), \ldots, z(t_n) - z(t_{n-1})$ are independent for $t_1 \leq t_2 \leq \cdots \leq t_n$.

Thus a vector Brownian motion is built upon the vector Gaussian random variable in the same way that a Brownian motion is built upon the one-dimensional Gaussian random variable. That is, increments of a vector Brownian motion are independent

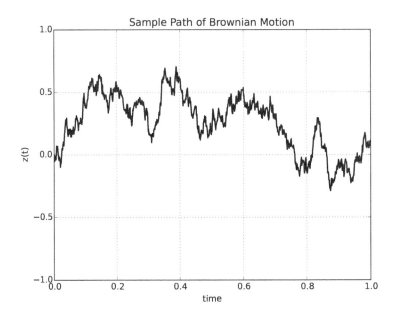

Figure 1.1 Typical sample path of Brownian motion.

and Gaussian distributed with mean 0 and a covariance matrix that is proportional to the time increment.

As we emphasized, Brownian motion has continuous sample paths that are too well behaved for some events we would like to model. For instance, market crashes, bankruptcy, etc. are often discontinuous price movements. Hence, we need a process that jumps! Poisson processes, which are built on the Poisson random variable, are what we are looking for.

1.1.3 Poisson Random Variables

A discrete random variable X taking values in the whole numbers is Poisson with parameter $\alpha > 0$ if

$$\mathbb{P}(X = k) = \frac{\alpha^k}{k!} e^{-\alpha}, \qquad k = 0, 1, \dots \qquad (1.5)$$

where $\mathbb{P}(X = k)$ is the probability that $X = k$. In this definition, note that we use the convention that $0! = 1$.

The mean of a Poisson random variable is $\mathbb{E}[X] = \alpha$ and the variance is $Var(X) = \sigma^2 = \alpha$. Further details on the Poisson random variable can be found in any standard probability textbook [19].

1.1.4   Poisson Processes

A Poisson process is a stochastic process built on the Poisson random variable as follows. A Poisson process with parameter α is a stochastic process $\pi(t; \alpha) : t \in \mathbb{R}^+$ that satisfies:

1. $\pi(0) = 0$.

2. $\pi(t) - \pi(s)$ is Poisson distributed with parameter $\alpha(t - s)$ for $t > s$.

3. $\pi(t_2) - \pi(t_1)$, $\pi(t_3) - \pi(t_2)$, $\ldots$, $\pi(t_n) - \pi(t_{n-1})$ are independent for $t_1 < t_2 \leq \cdots \leq t_n$.

Note that these three defining properties of the Poisson process mirror those used to define Brownian motion, with the Poisson random variable replacing the Gaussian.

For us, the most important property of a Poisson process is that it jumps! In fact, the Poisson process will often be used to tell us when jumps occur. Hence, it will play an important role in models that include market crashes, jumps, bankruptcies, and other unexpected discontinuous price movements. A typical sample path from a Poisson process with $\alpha = 1$ is given in Figure 1.2.

The parameter α is often called the *intensity* (or sometimes the *propensity*) of the Poisson process. You can think of it as the expected number of jumps in a single time period. Alternatively, you expect to see a single jump every $\frac{1}{\alpha}$ time periods. Therefore, the larger the intensity, the more frequent the jumps.

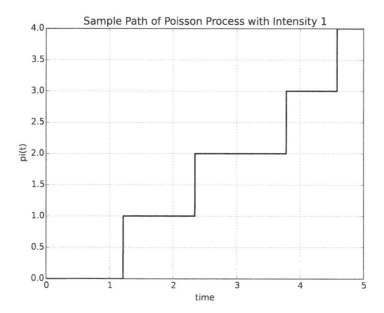

Figure 1.2 Sample path of a Poisson process.

As a technical detail, we will assume that a Poisson process is continuous from the right, and not the left. Thus, at the exact time that a Poisson process jumps, it takes on the new value to which it jumped. Functions that are right continuous and have left limits are often called *rcll* functions (or cadlag or R-functions). Because we have defined Poisson processes to take on the new value at a jump time, this makes the sample paths of Poisson processes *rcll* functions.

1.1.5 Increments of Brownian Motion and Poisson Processes

Over a small Δt, Brownian motion and Poisson processes can be thought of in simple and intuitive ways. First, we think of Δt as a "small" increment in t. When dealing with a stochastic process $X(t)$, we will also think of $\Delta X(t)$ as the change that occurs in X over a small time period Δt. That is,

$$\Delta X(t) = X(t + \Delta t) - X(t). \tag{1.6}$$

This notion of an increment of a stochastic process will guide our intuition. In this way, we can look at increments of Brownian motion and Poisson processes.

1.1.5.1 Brownian Motion Increment

Over a small time Δt, by the second defining property of Brownian motion, it is distributed as

$$\Delta z(t) = z(t + \Delta t) - z(t) \sim \mathcal{N}(0, \Delta t), \tag{1.7}$$

or written slightly differently as,

$$\Delta z(t) = Z\sqrt{\Delta t} \quad \text{where } Z \sim \mathcal{N}(0, 1). \tag{1.8}$$

That is, a Brownian motion increment looks like a standard (mean 0, variance 1) Gaussian multiplied by $\sqrt{\Delta t}$. Thus, it follows that $\mathbb{E}[\Delta z] = 0$ and $\mathbb{E}[(\Delta z)^2] = \Delta t$.

An even simpler picture arises by using a binary approximation to the standard normal random variable Z, given by

$$Z \approx \begin{cases} 1 & w.p. \quad 1/2 \\ -1 & w.p. \quad 1/2 \end{cases} \tag{1.9}$$

where *w.p.* stands for "with probability." Using this in place of Z in the formula for Δz yields the binary approximation,

$$\Delta z \approx \begin{cases} \sqrt{\Delta t} & w.p. \quad 1/2 \\ -\sqrt{\Delta t} & w.p. \quad 1/2. \end{cases} \tag{1.10}$$

This is depicted in Figure 1.3. This binary model of an increment of Brownian motion is useful for intuition and also for computation. In Chapter 9, computational schemes for pricing derivatives will rely heavily on this binary approximation.

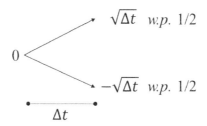

Figure 1.3 Binary model of an increment in Brownian motion.

1.1.5.2 Poisson Process Increment

A Poisson process can also be approximated over a small time period Δt. By the second defining property of a Poisson process, over Δt it appears as

$$\Delta \pi(t; \alpha) = \pi(t + \Delta t; \alpha) - \pi(t; \alpha) = X \quad \text{where} \quad X \sim Poisson(\alpha \Delta t). \quad (1.11)$$

Once again, there is a simple binary approximation which in the Poisson process case is given by

$$\Delta \pi(\alpha) \approx \begin{cases} 1 & w.p. & \alpha \Delta t \\ 0 & w.p. & 1 - \alpha \Delta t. \end{cases} \quad (1.12)$$

This binary model matches the mean and variance of $\Delta \pi(\alpha)$ to order Δt. A simple picture of this heuristic increment model is given in Figure 1.4. Note that for an

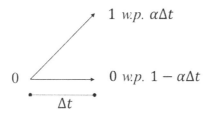

Figure 1.4 Binary model of an increment of a Poisson process.

actual Poisson process, $\Delta \pi$ takes the value 0 or 1 to order Δt, whereas all larger values of $\Delta \pi$ occur with probability of higher order in Δt. That is, roughly speaking, over "small" Δt, the Poisson process either doesn't jump, or jumps only once. More than one jump is a higher order occurrence.

Note the key difference between a Brownian motion and a Poisson process. From the simple binary model approximations, we see that in a Brownian motion the *size* of the move scales with the square root of Δt. Hence, over short periods of time, the move in a Brownian motion is also small. This is why Brownian motion has continuous paths. On the other hand, in a Poisson process, from the binary model we see that the move size can always be 1, regardless of how small Δt is. On the other hand, the *probability* of having a jump of size 1 scales with Δt and is small if

Δt is small. Thus, Poisson processes jump when they move. Over small periods of time, the probability of a jump is also small. This is the essential difference between Brownian motion and the Poisson process.

1.2 STOCHASTIC DIFFERENTIAL EQUATIONS

A simple way to think of a stochastic differential equation is as a differential equation driven by a stochastic process. We will use this point of view here. Also, to avoid the technicalities of stochastic calculus, we will present an intuitive approach to stochastic differential equations and stochastic differentials.

1.2.1 The Differential

Roughly speaking, the notion of a differential or infinitesimal of a process is that in an increment

$$\Delta X(t) = X(t + \Delta t) - X(t), \tag{1.13}$$

we can take Δt to be infinitesimally small. In such a case we would write

$$dX(t) = X(t + dt) - X(t) \tag{1.14}$$

where dt is "just a little bit of t." However, (1.14) has a problem when it comes to processes that jump and are assumed to be right continuous. The Poisson process is a good example.

1.2.1.1 The Problem with Jumps

We have to be very careful when a process has jumps and we assume right continuity of paths. Here is the problem. Assume that a Poisson process is currently at 0. That is $\pi(t) = 0$. Now, assume that a jump occurs at time s. Our convention will be to assume that at the exact time of the jump, the Poisson process takes the value 1. That is $\pi(s) = 1$. This means that Poisson processes and all other processes with jumps, will be continuous from the right. Hence,

$$\lim_{h \downarrow s} \pi(h) = \pi(s) \tag{1.15}$$

where $h \downarrow s$ denotes the limit from the right with $h > s$. This situation is shown in Figure 1.5.

Since this is our convention, let's consider defining the differential of a Poisson process as

$$d\pi(t) = \pi(t + dt) - \pi(t) \tag{1.16}$$

where $dt > 0$.

Now, we know that a jump occurred at time s, so intuitively the change in the process is equal to 1 at time s. That is, we should expect that $d\pi(s) = 1$. However, let us take the limit of $d\pi(t)$ for any t (including s) as $dt \downarrow 0$. We obtain

$$\lim_{dt \downarrow 0} d\pi(t) = \lim_{dt \downarrow 0} \pi(t + dt) - \pi(t) = \pi(t) - \pi(t) = 0, \tag{1.17}$$

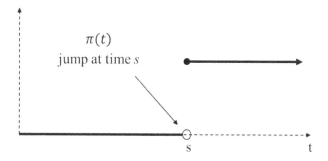

Figure 1.5 Poisson jump at time s.

where this calculation followed by right continuity as defined by equation (1.15). But this indicates that π never jumps since this equation holds for *every t*! Something must be wrong!

What is wrong is that we have assumed that π is right continuous, and then when we add dt to the current time, we are implicitly taking the limit from the right. Hence, we are guaranteed never to capture the jump!

This problem arises from our convention of assuming that Poisson processes are right continuous. If we had assumed left continuity, we would have avoided this specific issue. However, it is common in the literature to assume processes are right continuous, so we have adopted this convention. Therefore, we need to adjust our notion of a differential of a stochastic process slightly to account for this artifact.

1.2.1.2   Altering the Differential for Jumps

The solution to the above problem is that for a differential, we should think of the following

$$dX(t) = X(t + dt) - X(t^-) \qquad (1.18)$$

where $X(t^-) = \lim_{h \uparrow t} X(h)$ is the limit from the left of X at time t. By using the limit from the left, we make sure to capture jumps of the process, no matter how small dt is made.

We will develop this view because it provides the proper intuition. Hence, reviewing from above with Brownian motion we have,

$$dz(t) = z(t + dt) - z(t) \sim N(0, dt). \qquad (1.19)$$

Note that since Brownian motion has continuous sample paths, $z(t^-) = z(t)$. However, for a Poisson process, we should think of differentials as

$$d\pi(t; \alpha) = \pi(t + dt; \alpha) - \pi(t^-; \alpha) \sim Poisson(\alpha dt) \qquad (1.20)$$

in order to make sure that we capture jumps.

Don't forget that we also have the binary model approximations of Figures 1.3 and 1.4. Those binary models also provide the proper intuition, and in both cases, sums of them will limit at Brownian motion or a Poisson process. For the differential, we simply replace Δt by dt in (1.10) and (1.12), giving the approximations,

$$dz \approx \begin{cases} \sqrt{dt} & w.p. \quad 1/2 \\ -\sqrt{dt} & w.p. \quad 1/2 \end{cases} \tag{1.21}$$

and

$$d\pi(\alpha) \approx \begin{cases} 1 & w.p. \quad \alpha dt \\ 0 & w.p. \quad 1 - \alpha dt. \end{cases} \tag{1.22}$$

1.2.2 Compound Poisson Process

When Poisson processes jump, they jump up by 1. We can generalize this and allow the jump size to be random. Let $\pi(t;\alpha)$ be a Poisson process with jump times $t_1, t_2, \ldots$. Construct a new process $\pi^Y(t;\alpha)$, by assigning jump Y_1 at time t_1, Y_2 at time t_2, etc. where Y_1, Y_2, ... are independent and identically distributed (iid) random variables. This process can be written as

$$\pi^Y(t;\alpha) = \sum_{i=0}^{\pi(t;\alpha)} Y_i. \tag{1.23}$$

That is, at time t this process is the sum of $\pi(t;\alpha)$ iid copies of Y, where $\pi(t;\alpha)$ is a standard Poisson process. Processes of this form can also conveniently be written as integrals,

$$\pi^Y(t;\alpha) = \sum_{i=0}^{\pi(t;\alpha)} Y_i = \int_0^t Y_s d\pi(s;\alpha). \tag{1.24}$$

where $d\pi(s;\alpha) = 1$ at the jump times, and 0 otherwise. For this reason, we represent the differential form of a compound Poisson process by $Y d\pi(t;\alpha)$. That is, we may write

$$d\pi^Y = Y d\pi. \tag{1.25}$$

Following along the lines of the binary approximation to a Poisson process as in Figure 1.4, an infinitesimal approximation of a compound Poisson process can be thought of as

$$Y d\pi(\alpha) \approx \begin{cases} Y_i & w.p. \quad \alpha dt \\ 0 & w.p. \quad 1 - \alpha dt \end{cases} \tag{1.26}$$

and a heuristic infinitesimal picture of this is given in Figure 1.6. This binary approximation indicates that when a jump occurs, it is of size Y_i which is a random variable.

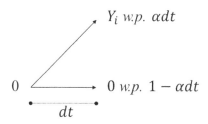

Figure 1.6 Infinitesimal model of a compound Poisson process.

1.2.3   Ito Stochastic Differential Equations

Stochastic integrals, and by extension stochastic differential equations, can be defined in different ways. The most useful for us is the Ito stochastic integral. At this point, I will not delve into the depths of the stochastic integral (because often people are never able to return!), but merely provide the intuition that you should take away when considering stochastic differential equations.

A stochastic differential equation will be written as

$$dx(t) = a(x(t), t)dt + b(x(t), t)dz(t) \qquad (1.27)$$

where in this case it is being driven by the Brownian motion $z(t)$. (At this stage, I will ignore the technical conditions that must be placed on a and b to make such an equation well defined. The interested reader should refer to a standard text on stochastic differential equations such as [42], [11], [34] or [44].) Appealing to our notion of the differential of a stochastic process, we will interpret this equation as

$$x(t + dt) - x(t) = a(x(t), t)dt + b(x(t), t)(z(t + dt) - z(t)). \qquad (1.28)$$

Since $z(t)$ has independent increments, and $a(x(t), t)$ and $b(x(t), t)$ are evaluated at time t, they are considered *independent* of the increment $dz(t) = z(t+dt) - z(t)$. This is important! It allows us to do the following simple calculations of the instantaneous mean, variance, and standard deviation of an Ito stochastic differential equation.

1.2.3.1   Instantaneous Statistics

Let $\mathbb{E}[\cdot|x]$ denote the conditional expectation given x. By the term "instantaneous statistics," we mean the expectation, variance, and standard deviation of dx given that we are at a specified state $x(t)$.

Assuming that the stochastic differential equation is given in equation (1.27), we can calculate the instantaneous mean $\mathbb{E}[dx|x(t)]$ as

$$
\begin{aligned}
\mathbb{E}[dx|x(t)] &= \mathbb{E}\left[a(x(t), t)dt + b(x(t), t)dz(t)|x(t)\right] \\
&= a(x(t), t)dt + b(x(t), t)\mathbb{E}\left[dz(t)|x(t)\right] \\
&= a(x(t), t)dt \qquad (1.29)
\end{aligned}
$$

where we used the fact that increments of Brownian motion have zero mean.

From this calculation, we see that $a(x(t), t)$ is the expected drift rate per unit time. Said another way, if time is measured in years, then $a(x(t), t)$ is the instantaneous mean expressed as an annualized quantity.

The instantaneous variance can be computed similarly as

$$
\begin{aligned}
\mathbb{E}[(dx - a(x(t), t)dt)^2 | x(t)] &= \mathbb{E}[b(x(t), t)^2 dz(t)^2 | x(t)] \\
&= b^2(x(t), t)\mathbb{E}[dz(t)^2 | x(t)] \\
&= b^2(x(t), t)dt
\end{aligned}
\tag{1.30}
$$

where we used the property of Brownian motion that $\mathbb{E}[dz(t)^2 | x(t)] = dt$. Thus, $b^2(x(t), t)$ represents the instantaneous variance rate. The instantaneous standard deviation is given by $b(x(t), t)\sqrt{dt}$.

Example 1.1 (Drift and Variance Rate of a Stochastic Differential Equation) Consider the Brownian motion driven stochastic differential equation,

$$
dx = 0.3x(t)dt + 0.4\sqrt{x(t)}dz,
$$

and assume that at time $t = 0.5$, the value is $x(0.5) = 10$. The instantaneous expected drift rate and variance rate are given by the formulas (1.29) and (1.30), respectively, evaluated at $x(0.5) = 10$. These lead to

$$
\text{drift rate} = a(x(t), t) = (0.3)x(0.5) = (0.3)(10) = 3,
$$

$$
\text{variance rate} = b^2(x(t), t) = (0.4)^2 x(0.5) = (0.4)^2(10) = 1.6.
$$

1.2.4 Poisson Driven Differential Equations

The same intuition can be used for Poisson driven differential equations. Consider the compound Poisson stochastic differential equation,

$$
dx(t) = a(x(t^-), t)dt + b(x(t^-), t)Y d\pi(t; \alpha)
\tag{1.31}
$$

where we have written $x(t^-)$ in the arguments of $a(\cdot)$ and $b(\cdot)$ to indicate that we are using limits from the left. We will assume that $a(\cdot)$ and $b(\cdot)$ are left continuous in the t argument so that we may use t instead of t^-. We will also sometimes use the notation x^- when we want to suppress the argument t, or even a^- when suppressing all the arguments of a.

The reason for using limits from the left was given in Section 1.2.1.1 where we argued that we would need to interpret our Poisson differential as

$$
d\pi(t) = \pi(t + dt) - \pi(t^-).
$$

For the Ito stochastic differential equation, we assume that the coefficients $a(\cdot)$ and $b(\cdot)$ are evaluated at the starting point of the differential time increment, which is t^-. This limit from the left is also important in the coefficients $a(\cdot)$ and $b(\cdot)$ because we want them to be independent of the Poisson increment $d\pi$. The only way we can do this is to make sure that we use left limits. This means that if $\pi(t)$ jumps at time t, which also causes a jump in $x(t)$ at time t, we use the value $x(t^-)$ in the argument of $a(\cdot)$ and $b(\cdot)$ so that they are evaluated *immediately* prior to the jump.

With these technicalities out of the way, we can compute the instantaneous mean as

$$
\begin{aligned}
\mathbb{E}[dx(t)|x(t^-)] &= \mathbb{E}[a(x(t^-),t)dt + b(x(t^-),t)Y\,d\pi(t)|x(t^-)] \\
&= a(x(t^-),t)dt + b(x(t^-),t)\mathbb{E}[Y\,d\pi(t)|x(t^-)] \\
&= \left(a(x(t^-),t) + b(x(t^-),t)\mathbb{E}[Y]\alpha\right)dt.
\end{aligned}
$$

In this case we see that the $d\pi(t)$ term can contribute to the instantaneous mean, and the expected instantaneous drift rate is $a(x(t^-),t) + b(x(t^-),t)\mathbb{E}[Y]\alpha$. This is because Poisson processes don't have zero mean, and this can make things messy! It is often better to think of the coefficient of dt (the first term in the stochastic differential equation) as the "mean" or "drift" term, and the coefficient of the driving stochastic process (the second term with $d\pi$ in this case) as a pure "risk" term without drift. To do this with a Poisson driven differential equation, we would like the second term to have zero instantaneous mean. Hence, we will often "compensate" the Poisson process to give it zero mean. This is done by simply subtracting the instantaneous mean from the second term and adding it to the first. It looks like

$$
dx(t) = \left[a(x(t^-),t) + b(x(t^-),t)\mathbb{E}[Y]\alpha\right]dt + b(x(t^-),t)\left[Y\,d\pi(t) - \mathbb{E}[Y]\alpha dt\right].
$$

We can also compute the instantaneous variance as

$$
\begin{aligned}
&\mathbb{E}[(dx(t) - \left[a(x(t^-),t) + b(x(t^-),t)\mathbb{E}[Y]\alpha\right]dt)^2\,|x(t^-)] \\
&\quad = \mathbb{E}[b^2(x(t^-),t)(Y\,d\pi(t) - \mathbb{E}[Y]\alpha dt)^2|x(t^-)] \\
&\quad = b^2(x(t^-),t)Var(Y\,d\pi(t)) \\
&\quad = b^2(x(t^-),t)\left(\mathbb{E}[Y^2 d\pi(t)^2] - \mathbb{E}[Y]^2\mathbb{E}[d\pi(t)]^2\right) \\
&\quad = b^2(x(t^-),t)\left(\mathbb{E}[Y^2]\mathbb{E}[d\pi(t)^2] - \mathbb{E}[Y]^2\alpha^2 dt^2\right) \\
&\quad = b^2(x(t^-),t)\left(\mathbb{E}[Y^2](\alpha dt + \alpha^2 dt^2) - \mathbb{E}[Y]^2\alpha^2 dt^2\right) \\
&\quad = b^2(x(t^-),t)\mathbb{E}[Y^2]\alpha dt + O(dt^2)
\end{aligned}
$$

where $Var(Y\,d\pi(t))$ was used to denote the variance of $Y\,d\pi(t)$, and we used the identity $Var(X) = \mathbb{E}[X^2] - (\mathbb{E}[X])^2$. Hence, to order dt, the instantaneous variance rate is given by $b^2(x(t^-),t)\mathbb{E}[Y^2]\alpha$.

Example 1.2 (Drift and Variance Rate under Poisson)
Consider the compound Poisson driven stochastic differential equation,

$$dx = 2(x(t^-) - t)dt + 0.1x(t^-)Y d\pi(t; 0.8),$$

where $\mathbb{E}[Y] = 0.05$ and $Var(Y) = 0.1$. The instantaneous expected drift and variance rates at time $t = 0.25$ and $x(0.25^-) = 1.5$ are given by

$$
\begin{aligned}
\text{drift rate} &= 2(x(t^-) - t) + (0.1)x(t^-)\mathbb{E}[Y]\alpha \\
&= 2(1.5 - 0.25) + (0.1)(1.5)(0.05)(0.8) = 2.506,
\end{aligned}
$$

$$
\begin{aligned}
\text{variance rate} &= (0.1^2)(x^2(t^-))\mathbb{E}[Y^2]\alpha \\
&= (0.1^2)(x^2(t^-))\left(Var(Y) + (\mathbb{E}[Y])^2\right)\alpha \\
&= (0.1^2)(1.5)^2(0.1 + 0.05^2)(0.8) = 0.00185.
\end{aligned}
$$

1.3 SUMMARY

Brownian motion (built upon the Gaussian random variable) and the Poisson process (built upon the Poisson random variable) are the basic building blocks used to create models of prices and financial variables. In particular, we use these two processes to drive differential equations, and this allows us to capture a wide range of phenomena. Due to the continuity of Brownian motion, it is good for modeling price paths and variables that do not jump. On the other hand, Poisson processes are essential building blocks for modeling jumps in price processes or variables.

Much intuition can be gained from simple "incremental" and "differential" models of processes and stochastic differential equations. The simple binary approximations to Brownian motion and Poisson processes are enough to correctly guide your intuition in the vast majority of cases. Thus, for modeling purposes, make sure you have a solid understanding of these two building block processes.

EXERCISES

1.1 Calculate the instantaneous expected drift rate and variance rate of the following Brownian motion driven stochastic differential equations at time $t = 0.5$ and $x(0.5) = 2$:

(a) $dx = 3(1 - x(t))dt + 0.4dz.$

(b) $dx = 0.1x(t)dt + 0.2x(t)dz.$

(c) $dx = 2t^2 x(t)dt + 0.3t\sqrt{x(t)}dz.$

1.2 Calculate the instantaneous expected drift rate and variance rate of the following Poisson stochastic differential equations at $t = 0.5$ and $x(0.5^-) = 2$ where Y is a random variable with mean $\mathbb{E}[Y] = 0.1$ and $Var(Y) = 0.04$:

(a) $dx = 3(1 - x(t^-))dt + 0.4d\pi(t; 0.3)$.

(b) $dx = 0.1x(t^-)dt + 0.2x(t^-)Y d\pi(t; 2)$.

(c) $dx = 0.3tx(t^-)dt + 0.2(x(t^-) + t)Y d\pi(t; 0.6)$.

1.3 Verify that our binary model of a Poisson process over time dt,

$$d\pi = \begin{cases} 1 & w.p. & \alpha dt \\ 0 & w.p. & 1 - \alpha dt \end{cases} \tag{P1.1}$$

has a mean and variance that agree with a Poisson random variable with parameter αdt to order dt.

1.4 Poisson Processes.

Consider the time interval $[0, 1]$. Chop this time interval into n parts of equal length. Over each interval define the independent and identically distributed random variables X_i where

$$X_i = \begin{cases} 1 & w.p. & \alpha/n \\ 0 & w.p. & 1 - \alpha/n. \end{cases} \tag{P1.2}$$

Let

$$Y = \sum_{i=1}^{n} X_i. \tag{P1.3}$$

Answer the following, where $\mathbb{P}(\cdot)$ denotes probability.

(a) What is $\mathbb{P}(Y = 0)$?

(b) In your answer in (a), take the limit as $n \to \infty$. What do you get?

(b) What is $\mathbb{P}(Y = 1)$?

(c) Again take the limit. What is your answer?

(d) Now consider an arbitrary but fixed k with $k < n$. What is $\mathbb{P}(Y = k)$?

(e) Again take the limit as $n \to \infty$, and show that this converges to the Poisson random variable. (You will probably want to use Stirling's formula $n! \sim \sqrt{2\pi} e^{-n} n^{n+\frac{1}{2}}$. This calculation is a bit tricky!)

(Note: In this problem we converge to a Poisson random variable with parameter α since we took the time interval to be 1. If the time interval is t, we will converge to a Poisson random variable with parameter αt. As a function of t, we arrive at a Poisson process.)

1.5 Poisson Processes Again.

Consider the following Markov chain. Let the state space be the whole numbers

$x = 0, 1, 2, ...$ Consider the following transition probabilities over the time instant dt:

$$\mathbb{P}(x(t+dt) = n | x(t) = n) = 1 - \alpha dt. \qquad (P1.4)$$
$$\mathbb{P}(x(t+dt) = n+1 | x(t) = n) = \alpha dt. \qquad (P1.5)$$

Notationally, let $p_n(t) = \mathbb{P}(x(t) = n)$.

(a) Write a differential equation for $p_0(t)$. (Hint: to derive a differential equation, consider the amount of probability that flows into and out of the state $x = 0$ over time dt.)

(b) Assume $p_0(0) = 1$ (that is, at time zero, $x = 0$ with probability 1). Solve the differential equation for $p_0(t)$.

(c) Derive a differential equation for $p_n(t)$, $n > 0$. Given your answer in (a), solve for $p_1(t)$. Explain how you could solve for $p_n(t)$ for any n.

(Note: Again we have arrived at a Poisson process, but this time through Markov chain theory. A Poisson process is an example of a continuous time Markov process, and the set of differential equations you derived is the "forward equation" for this process. See, for example, [21])

1.6 Simple Formula for the Variance.

Recall that the variance of a random variable is defined as

$$Var(X) = \mathbb{E}[(X - \mathbb{E}[X])^2].$$

Show that this formula for the variance can also be expressed as

$$Var(X) = \mathbb{E}[X^2] - (\mathbb{E}[X])^2.$$

1.7 Time Scaling of the Variance and Independent Increments.

(a) Show that for two independent random variables, X and Y, the variance of the sum is the sum of the variances. That is, show that

$$Var(X + Y) = Var(X) + Var(Y)$$

if X and Y are independent.

(b) Consider a stochastic process with independent increments where all increments of a fixed time length are identically distributed. Use the result of part (a) to show that that the variance of a process with independent and identically distributed increments must be proportional to time t. More specifically, let N be a specified number of equal time increments from 0 to T with $\Delta t = \frac{T}{N}$. Also define $\Delta X_i = X((i+1)\Delta t) - X(i\Delta t)$ to be independent and identically distributed increments of the stochastic process $X(t)$. Show that

$$Var(X(T)) \propto T.$$

1.8 Correlating Random Variables.

Let Y be a vector random variable with mean 0 and covariance matrix Σ. Furthermore, let L be the so-called *Cholesky* factorization of Σ such that $LL^T = \Sigma$. If X is a vector Gaussian random vector with mean vector 0 and identity covariance matrix $\mathbb{E}[XX^T] = I$, show that $Y = LX$ has covariance matrix equal to Σ. This indicates a simple method of constructing a random vector with a specific covariance structure from a random vector with identity covariance structure.

1.9 Mean and Variance of the Poisson Random Variable.

Let X be a Poisson random variable with parameter α. Show that $\mathbb{E}[X] = \alpha$ and $Var(X) = \alpha$.

1.10 Exponential Waiting Times and Poisson Processes.

Let $\pi(t)$ be a Poisson process with intensity α. Show that the waiting time to the first jump is exponentially distributed. That is, let t^* be the time at which $\pi(t)$ jumps from 0 to 1. Show that the cumulative distribution function $F_{t^*}(T) = \mathbb{P}(t^* \leq T) = 1 - e^{-\alpha T}$ and that the density function is $f_{t^*}(t) = \frac{dF_{t^*}(t)}{dt} = \alpha e^{-\alpha t}$.

CHAPTER 2

Ito's Lemma

I TO'S lemma is the chain rule for stochastic calculus. In this chapter we present Ito's lemma for Brownian motion and Poisson process driven stochastic differential equations. It has been said that (almost) all of mathematical finance can be done with just the knowledge of Ito's lemma. This means that you should make sure that you know (and understand) Ito's lemma.

2.1   CHAIN RULE OF ORDINARY CALCULUS

In ordinary calculus, here is how the chain rule works in conjunction with a differential equation. Let $x(t) \in \mathbb{R}$ follow the ordinary differential equation

$$\frac{dx}{dt} = a(x,t). \tag{2.1}$$

Now consider a function of $x(t)$ and t. Let's call this function $f(x(t),t)$. Assuming that f is differentiable, we can ask what the time derivative of f is. To calculate it, we simply apply the chain rule,

$$\frac{df(x,t)}{dt} = \frac{\partial f}{\partial x}\frac{dx}{dt} + \frac{\partial f}{\partial t} = f_x \frac{dx}{dt} + f_t \tag{2.2}$$

where we are using the notation $f_x = \frac{\partial f}{\partial x}$ and $f_t = \frac{\partial f}{\partial t}$. Finally, we can substitute for $\frac{dx}{dt}$ in (2.1), giving

$$\frac{df(x,t)}{dt} = \frac{\partial f}{\partial x}a(x,t) + \frac{\partial f}{\partial t} = a(x,t)f_x + f_t. \tag{2.3}$$

This is a fairly straightforward calculation. However, when dealing with stochastic differential equations, the simple chain rule of ordinary calculus does not work. The reason is simple. Brownian motion is not differentiable so we can't really take its derivative or the derivative of any function of Brownian motion. Also, Poisson processes jump, and at these jumps they are not even continuous, let alone differentiable. Thus, for stochastic differential equations we need to develop the correct mathematics for dealing with a function of a variable that follows a stochastic differential equation. The guiding result is known as Ito's lemma.

2.1.1 Multivariable Taylor Series Expansions

Before diving into Ito's lemma, you should make sure that you recall your multivariable Taylor series expansions to second order (see any standard calculus reference, such as [48]). This is extremely important! Ito's lemma for Brownian motion is basically just a modified Taylor expansion to second order. So, let's recall up to second order the Taylor series expansion of a function $f(x,t)$ around a point (x_0, t_0),

$$
\begin{aligned}
f(x,t) \ = \ & f(x_0, t_0) + f_t(x_0, t_0)(t - t_0) + f_x(x_0, t_0)(x - x_0) \\
& + \frac{1}{2} f_{tt}(x_0, t_0)(t - t_0)^2 + f_{xt}(x_0, t_0)(x - x_0)(t - t_0) \\
& + \frac{1}{2} f_{xx}(x_0, t_0)(x - x_0)^2 + \dots
\end{aligned}
\tag{2.4}
$$

Now, we will typically denote $dx = x - x_0$, $dt = t - t_0$, and $df = f(x,t) - f(x_0, t_0)$, so that a Taylor series expansion can be written as

$$
df \ = \ f_t dt + f_x dx + \frac{1}{2} f_{tt} dt^2 + f_{xt} dx dt + \frac{1}{2} f_{xx} dx^2 + \dots
\tag{2.5}
$$

where the arguments of the partial derivatives, f_t, f_x, etc., have been suppressed.

In the multivariable case when $x \in \mathbb{R}^n$, the Taylor series expansion is

$$
df \ = \ f_t dt + f_x dx + \frac{1}{2} f_{tt} dt^2 + dx^T f_{xt} dt + \frac{1}{2} dx^T f_{xx} dx + \dots
\tag{2.6}
$$

where $f_x \in \mathbb{R}^{1 \times n}$ is the gradient, $f_{xt} \in \mathbb{R}^{n \times 1}$ is a column vector of mixed partial derivatives, and $f_{xx} \in \mathbb{R}^{n \times n}$ is the Hessian matrix, defined as

$$
f_x = [f_{x_1}, \dots, f_{x_n}], \quad f_{xt} = \begin{bmatrix} f_{x_1 t} \\ \vdots \\ f_{x_n t} \end{bmatrix}, \quad f_{xx} = \begin{bmatrix} f_{x_1 x_1} & \cdots & f_{x_1 x_n} \\ \vdots & \ddots & \vdots \\ f_{x_n x_1} & \cdots & f_{x_n x_n} \end{bmatrix}.
\tag{2.7}
$$

A useful trick in the multivariable case is to note that the last term in equation (2.6) can be written as

$$
dx^T f_{xx} dx = Tr(dx^T f_{xx} dx) = Tr(f_{xx} dx dx^T),
\tag{2.8}
$$

where $Tr(\cdot)$ is the trace of a matrix (i.e., the sum of the diagonal elements [49]).

Operationally, Ito's lemma for Brownian motion boils down to nothing more than substituting $dx = a dt + b dz$ in the Taylor expansion of (2.5) (or (2.6) in the multivariable case), interpreting terms containing dz correctly, and then throwing away terms of order higher than dt.

If that sounds simple, you are right. Let's see how it works in more detail.

2.2 ITO'S LEMMA FOR BROWNIAN MOTION

Given the differential of $x(t)$, Ito's lemma allows us to compute the differential of a function of $x(t)$ and t. Hence, it is the "chain rule" for stochastic differential equations. The following result is Ito's lemma when $x(t) \in \mathbb{R}$ is a process governed by a stochastic differential equation driven by a scalar Brownian motion $z(t) \in \mathbb{R}$.

(★) **Ito's Lemma for Brownian Motion**
Consider the stochastic differential equation,

$$dx = a(x,t)dt + b(x,t)dz \tag{2.9}$$

and let $f(x,t)$ be a twice continuously differentiable function of x and t. Then

$$df(x,t) = \left(f_t + a(x,t)f_x + \frac{1}{2}b^2(x,t)f_{xx} \right) dt + b(x,t)f_x dz. \tag{2.10}$$

Rigorous justification of Ito's lemma, which was introduced in [31], can be found in many references, such as [11, 34, 42]. Here I provide the following heuristic derivation.

Heuristic Proof: I will suppress the arguments of a and b for convenience. Consider writing the Taylor expansion of df,

$$
\begin{aligned}
df &= f(x(t+dt), t+dt) - f(x(t), t) \\
&= f_t dt + f_x dx + \frac{1}{2} f_{tt}(dt)^2 + \frac{1}{2} f_{xx}(dx)^2 + f_{xt} dx dt + ...
\end{aligned}
$$

Next, substitute in for dx using $dx = adt + bdz$,

$$
\begin{aligned}
df &= f_t dt + f_x(adt + bdz) + \frac{1}{2} f_{tt}(dt)^2 \\
&\quad + \frac{1}{2} f_{xx}(adt + bdz)^2 + f_{xt}(adt + bdz)dt + ... \\
&= f_t dt + f_x adt + f_x bdz + \frac{1}{2} f_{tt}(dt)^2 \\
&\quad + \frac{1}{2} f_{xx}(a^2 dt^2 + 2abdtdz + b^2 dz^2) + f_{xt}(adt^2 + bdzdt) + ...
\end{aligned}
$$

Now we take a crucial step and only keep terms up to order dt using the following logic. Recall that the standard deviation of dz is equal to $\sqrt{dt}$, and thus we think of the average size of dz as being $dt^{1/2}$. Moreover, this also implies that dz^2 is of order dt. Using this logic and keeping only terms up to order dt yields

$$df = f_t dt + f_x adt + f_x bdz + \frac{1}{2} f_{xx} b^2 dz^2.$$

Finally we replace dz^2 by its expectation dt (I haven't justified why this is reasonable) which leads to Ito's lemma,

$$df = (f_t + af_x + \frac{1}{2}b^2 f_{xx})dt + bf_x dz.$$

□

In this "derivation" there were a couple of dubious steps. The most glaring was replacing dz^2 by its expectation dt. Let's see why this was a reasonable thing to do.

2.2.1   Replacing dz^2 by dt

Here is a simple argument as to why it is reasonable to replace dz^2 by dt. If we can say that $dz^2 = dt$, integrating this would imply

$$\int_0^T dz^2 = \int_0^T dt = T. \tag{2.11}$$

Therefore, let us see if this makes sense. Let's approximate the integral above by the sum

$$\mathcal{S}(T, \Delta t) = \sum_{i=0}^{\frac{T}{\Delta t}-1} (z((i+1)\Delta t) - z(i\Delta t))^2 \approx \int_0^T dz^2. \tag{2.12}$$

Now, the claim is that as $\Delta t \to 0$, then the sum $\mathcal{S}(T, \Delta t) \to T$. But note that $\mathcal{S}(T, \Delta t)$ is a random variable since it involves $z(t)$. Therefore, we are trying to show that the random variable $\mathcal{S}(T, \Delta t)$ converges to the constant T. To do this, we will show that the mean of $\mathcal{S}(T, \Delta t)$ is equal to T, and its variance approaches zero. This does the trick, since a random variable with zero variance must be a constant equal to its mean. (When we show that the variance approaches zero, we are proving convergence in mean square, or $L^2(\mathbb{P})$.)

2.2.1.1   Computing the Mean

Let's first compute the mean of $\mathcal{S}(T, \Delta t)$:

$$
\begin{aligned}
\mathbb{E}[\mathcal{S}(T, \Delta t)] &= \mathbb{E}\left[\sum_{i=0}^{\frac{T}{\Delta t}-1} (z((i+1)\Delta t) - z(i\Delta t))^2\right] \\
&= \sum_{i=0}^{\frac{T}{\Delta t}-1} E\left[(z((i+1)\Delta t) - z(i\Delta t))^2\right] \\
&= \sum_{i=0}^{\frac{T}{\Delta t}-1} \Delta t = T. \tag{2.13}
\end{aligned}
$$

Hence, the mean is indeed T.

2.2.1.2 Computing the Variance

Now let's compute the variance of $\mathcal{S}(T, \Delta t)$. By independent increments

$$Var(\mathcal{S}(T), \Delta t) \quad = \quad Var\left(\sum_{i=0}^{\frac{T}{\Delta t}-1}(z((i+1)\Delta t) - z(i\Delta t))^2\right)$$

$$= \quad \sum_{i=0}^{\frac{T}{\Delta t}-1} Var((z((i+1)\Delta t) - z(i\Delta t))^2). \qquad (2.14)$$

But since $z((i+1)\Delta t) - z(i\Delta t)$ is Gaussian with mean zero and variance Δt, we have

$$Var((z((i+1)\Delta t) - z(i\Delta t))^2) \quad = \quad \mathbb{E}[(z((i+1)\Delta t) - z(i\Delta t))^4]$$
$$- \mathbb{E}[(z((i+1)\Delta t) - z(i\Delta t))^2]^2$$
$$= \quad 3(\Delta t)^2 - (\Delta t)^2$$
$$= \quad 2(\Delta t)^2.$$

Therefore as $\Delta t \to 0$,

$$Var(\mathcal{S}(T, \Delta t)) \quad = \quad \sum_{i=0}^{\frac{T}{\Delta t}-1} Var((z((i+1)\Delta t) - z(i\Delta t))^2)$$

$$= \quad \sum_{i=0}^{\frac{T}{\Delta t}-1} 2(\Delta t)^2 = 2T\Delta t \to 0.$$

Hence, the limit of the variance of $\mathcal{S}(T, \Delta t)$ is zero. That means that the limit is a constant and equal to the mean T. Note that the above argument has much of the flavor of the weak law of large numbers [15].

This is the essential argument that allows us to use $dz^2 = dt$ and a simplified version of the argument behind a real derivation of Ito's lemma. Let's see an example of Ito's lemma applied to so-called geometric Brownian motion.

Example 2.1 (Geometric Brownian Motion)
Let $dx = axdt + bxdz$ where a and b are constants, and consider $f(x) = \ln(x)$. Then, according to Ito's lemma, $f(x)$ satisfies

$$df \quad = \quad (f_t + axf_x + \frac{1}{2}b^2x^2f_{xx})dt + bxf_xdz \qquad (2.15)$$

$$= \quad (a - \frac{1}{2}b^2)dt + bdz \qquad (2.16)$$

where $f_t = 0$, $f_x = \frac{1}{x}$, and $f_{xx} = -\frac{1}{x^2}$.

2.2.2 Discussion of Ito's Lemma

At the risk of overdoing an attempt to provide intuition behind Ito's lemma, I will leave you with the following thoughts.

In Ito's lemma, we kept terms up to order dt in the Taylor expansion. We are used to doing things like this from ordinary calculus, but now that we are dealing with stochastic processes, we might question this step. In particular, the standard deviation of dz is $dt^{1/2}$. This means that moves in dz are usually much *larger* than dt. Why doesn't this dz term completely dominate and even allow us to ignore terms of order dt?

To explain this, I am going to appeal to the wisdom of Daniel T. Gillespie (private communication, circa 2000). He gave the following explanation, which is based on the story of the tortoise and the hare. As the story goes, the tortoise and the hare race each other. The tortoise being slow, starts the race and steadily works his way toward the finish line. The hare, on the other hand, is quick and jumpy. He is much faster than the hare, but runs forward and backwards and easily gets off track. In the end of the story, the tortoise wins the race.

Now you're asking, "What does this have to do with Ito's lemma and stochastic differential equations?" Well, the deterministic dt drift term is like the tortoise. It marches forward at a constant dt rate. On the other hand, the dz term is like the hare. It is quick and jumps around like order $dt^{1/2}$. But its direction is random. Some of the time it jumps forward and at other times, backwards. Together, both the dt and dz terms contribute to the stochastic differential equation and neither term is guaranteed to dominate, just as we don't know whether the tortoise or the hare will win the race!

2.3 ITO'S LEMMA FOR POISSON PROCESSES

Ito's lemma for Brownian motion is more subtle than Ito's lemma for Poisson processes. (The key difference is that Poisson processes have sample paths of finite variation, and this allows us to define the stochastic integral pathwise. Hence, Ito's lemma becomes an application of the so-called Lebesgue–Stieltjes calculus [45].)

(★) Ito's Lemma for Poisson Processes
Given a Poisson stochastic differential equation (SDE)

$$dx = a(x^-, t)dt + b(x^-, t)d\pi \qquad (2.17)$$

where $x^- = x(t^-)$, let $f(x, t)$ be a continuously differentiable function of x and t. Then

$$df(x, t) = (f_t + a(x^-, t)f_x)dt + (f(x^- + b(x^-, t), t) - f(x^-, t))d\pi, \qquad (2.18)$$

where the partial derivatives of f are also evaluated at x^- and t.

Once again I will provide a simple heuristic derivation (based on [52]). This time when we consider the differential df we have to be careful because $x(t)$ jumps! Taylor expansions work well as approximations when dx is small; however with jumps dx can be large. Note how this plays into our derivation.

Heuristic Proof: We start by writing the differential and substituting for dx,

$$
\begin{aligned}
df &= f(x(t+dt), t+dt) - f(x^-, t) \\
&= f(x^- + a^- dt + b^- d\pi, t+dt) - f(x^-, t),
\end{aligned}
$$

where $x^- = x(t^-)$, $a^- = a(x(t^-), t)$ and $b^- = b(x(t^-), t)$. Now, we note that possible jumps come from $d\pi$. I don't like this term, so I will add and subtract a term that doesn't contain the jump,

$$
\begin{aligned}
df &= f(x^- + a^- dt + b^- d\pi, t+dt) - f(x^- + a^- dt, t+dt) \\
&\quad + f(x^- + a^- dt, t+dt) - f(x^-, t). \qquad (2.19)
\end{aligned}
$$

The final two terms don't contain jumps, so I can approximate their difference using ordinary calculus,

$$
\begin{aligned}
df &= f(x^- + a^- dt + b^- d\pi, t+dt) - f(x^- + a^- dt, t+dt) \\
&\quad + (f_t + a^- f_x) dt + O(dt^2), \qquad (2.20)
\end{aligned}
$$

where $O(dt^2)$ contains terms of order dt^2 and higher. Next, let's analyze the first two terms

$$
f(x^- + a^- dt + b^- d\pi, t+dt) - f(x^- + a^- dt, t+dt). \qquad (2.21)
$$

When there is no jump in $d\pi$, this term is zero. Assume that the intensity of the Poisson process is α, so that no jumps occur with probability $1 - \alpha dt$. When there is a jump, we have $d\pi = 1$ and

$$
f(x^- + a^- dt + b^-, t+dt) - f(x^- + a^- dt, t+dt). \qquad (2.22)
$$

This occurs with probability αdt. Since the $a^- dt$ term in the argument is of order dt, when combined with the probability of the jump, αdt, the overall effect of the $a^- dt$ term is of order dt^2 and can be ignored. Therefore, as $dt \to 0$, to order dt this entire term can be replaced by

$$
f(x^- + a^- dt + b^-, t+dt) - f(x^- + a^- dt, t+dt) \to f(x^- + b^-, t) - f(x^-, t).
$$

Hence, combining the above arguments gives

$$
f(x^- + a^- dt + b^- d\pi, t+dt) - f(x^- + a^- dt, t+dt) \to (f(x^- + b^-, t) - f(x^-, t)) d\pi.
$$

Substituting this in (2.20) completes the derivation of Ito's lemma in the Poisson case. □

2.3.1 Interpretation of Ito's Lemma for Poisson

Ito's lemma for Poisson processes simply says that when the Poisson process doesn't jump, use ordinary calculus. When it jumps, the move in f is determined completely by the jump. That is it! Let's work through an example of this.

Example 2.2 (Geometric Poisson Motion)
Let $dx = ax^- dt + (b-1)x^- d\pi$ where a and b are constant, and consider $f(x) = \ln(x)$. Note that this is the Poisson equivalent of geometric Brownian motion. By Ito's lemma for Poisson processes, $f(x)$ satisfies

$$
\begin{aligned}
df &= (f_t + f_x ax^-)dt + (f(x^- + (b-1)x^-) - f(x^-))d\pi \\
&= a\,dt + (\ln(bx^-) - \ln(x^-))d\pi \\
&= a\,dt + \ln(b)d\pi
\end{aligned}
$$

where $f_t = 0$ and $f_x = \frac{1}{x}$.

2.4 MORE VERSIONS OF ITO'S LEMMA

In this section I will present other useful versions of Ito's lemma. Heuristic derivations follow along the lines of those for Brownian motion and the Poisson process. For a rigorous treatment that encompasses all these versions of Ito's lemma, see [44, 45].

2.4.1 Ito's Lemma for Compound Poisson Processes

When we are using a compound Poisson process, Ito's lemma is modified slightly as follows.

(★) Ito's Lemma for Compound Poisson Processes
Given a compound Poisson driven stochastic differential equation,

$$ dx = a(x^-, t)dt + b(x^-, t)Y\,d\pi, \tag{2.23} $$

let $f(x, t)$ be a continuously differentiable function of x and t. Then

$$ df(x, t) = (f_t + a^- f_x)dt + (f(x^- + Yb^-, t) - f(x^-, t))d\pi, \tag{2.24} $$

where it is assumed that partial derivatives of f are evaluated at x^- and t.

This version of Ito's lemma can be justified in a manner similar to Ito's lemma for Poisson processes.

2.4.2 Ito's Lemma for Brownian and Compound Poisson

If we combine Brownian motion and a compound Poisson process, we have the following result.

(★) Ito's Lemma for Brownian Motion and Poisson Processes

Given the following stochastic differential equation driven by both a Brownian motion and a compound Poisson process,

$$dx = a(x^-, t)dt + b(x^-, t)dz + Y d\pi, \qquad (2.25)$$

let $f(x, t)$ be a twice continuously differentiable function of x and t. Then

$$df(x, t) = \left(f_t + a(x^-, t)f_x + \frac{1}{2}b(x^-, t)^2 f_{xx} \right) dt + b(x^-, t)f_x dz$$
$$+ \left(f(x^- + Y, t) - f(x^-, t) \right) d\pi. \qquad (2.26)$$

2.4.3 Ito's Lemma for Vector Processes

If we have multiple Brownian motions driving a stochastic differential equation, we have the following results.

(★) Ito's Lemma for Two Correlated Brownians

Consider the stochastic differential equations

$$dx_1 = a_1 dt + b_1 dz_1, \qquad (2.27)$$
$$dx_2 = a_2 dt + b_2 dz_2, \qquad (2.28)$$

with $a_1 = a_1(x_1, x_2, t)$, $a_2 = a_2(x_1, x_2, t)$, $b_1 = b_1(x_1, x_2, t)$, $b_2 = b_2(x_1, x_2, t)$ where z_1 and z_2 are two correlated Brownian motions with instantaneous correlation coefficient ρ (i.e., $\mathbb{E}[dz_1 dz_2] = \rho dt$). Let $f(x_1, x_2, t)$ be a twice continuously differentiable function of x_1, x_2 and t. Then

$$df(x_1, x_2, t) = \left(f_t + a_1 f_{x_1} + a_2 f_{x_2} + \frac{1}{2}b_1^2 f_{x_1 x_1} + \frac{1}{2}b_2^2 f_{x_2 x_2} + \rho b_1 b_2 f_{x_1 x_2} \right) dt$$
$$+ b_1 f_{x_1} dz_1 + b_2 f_{x_2} dz_2. \qquad (2.29)$$

Using vector notation we can generalize the above result, which leads to Ito's lemma for functions of vector processes driven by vector Brownian motion.

(★) Ito's Lemma for Vector Processes

Consider a vector stochastic differential equation

$$dx = adt + Bdz \qquad (2.30)$$

with $x \in \mathbb{R}^n$, $a = a(x,t) \in \mathbb{R}^n$, $B = B(x,t) \in \mathbb{R}^{n \times m}$ and $z \in \mathbb{R}^m$ a vector Brownian motion with instantaneous covariance structure $\mathbb{E}[dzdz^T] = \Sigma dt$. Let $f(x,t)$ be a twice continuously differentiable function of x and t. Then

$$df = \left(f_t + f_x a + \frac{1}{2} Tr(f_{xx} B \Sigma B^T) \right) dt + f_x B dz \qquad (2.31)$$

where $f_x \in \mathbb{R}^{1 \times n}$ is the gradient and $f_{xx} \in \mathbb{R}^{n \times n}$ is the Hessian matrix, defined respectively as

$$f_x = [f_{x_1}, f_{x_2}, ..., f_{x_n}] \quad and \quad f_{xx} = \begin{bmatrix} f_{x_1 x_1} & \cdots & f_{x_1 x_n} \\ \vdots & \ddots & \vdots \\ f_{x_n x_1} & \cdots & f_{x_n x_n} \end{bmatrix}. \qquad (2.32)$$

2.5 ITO'S LEMMA, THE PRODUCT RULE, AND A RECTANGLE

In Ito's lemma for Brownian motion, the mysterious term, $\frac{1}{2}b^2 f_{xx} dt$, enters because we are forced to keep the dz^2 term which is of order dt. Without this term, Ito's lemma would follow from the intuition of ordinary calculus.

Let me give a little argument (which appears in Rogers and Williams [45]) to try to convince you that Ito's lemma might actually be more intuitive than ordinary calculus. The example I will use is the product rule. In ordinary calculus, we have the familiar formula

$$d(uv) = udv + vdu. \qquad (2.33)$$

In fact, this is the beginning of the integration by parts formula. Let's try to "derive" this formula by a simple "rectangle" argument. Here it is.

Consider the quantity $d(uv)$. This is a change in the product uv when we change u and v by some small amounts. That is,

$$d(uv) = (u + du)(v + dv) - uv. \qquad (2.34)$$

We can think of this as the area of a rectangle with sides of length $u + du$ and $v + dv$, minus the area of a rectangle with sides of u and v. The rectangle in Figure 2.1 shows this. Now, it should be obvious that the area of $d(uv)$ is equal to the sum of the three rectangles contained inside the dotted lines in the figure. That is,

$$d(uv) = udv + vdu + dudv. \qquad (2.35)$$

Hence, it is natural to expect to see a term related to $dudv$! In fact, it is more

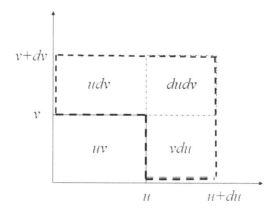

Figure 2.1 Ito's rectangle.

mysterious that in the ordinary calculus version, equation (2.33), we are able to ignore this term. Of course, that is because in ordinary calculus, roughly speaking, the terms du and dv are of order dt and hence $dudv$ is a higher order term.

Let us see how far we can get with this simple rectangle. Take $u = v = z(t)$. Then our rectangle formula (product rule) says that

$$d(z^2) = d(z \cdot z) = zdz + zdz + dz^2. \tag{2.36}$$

Note that the formula above is exact! Now, recalling that dz^2 should be replaced by dt as in the argument of Section 2.2.1, we have

$$d(z^2) = 2zdz + dt \tag{2.37}$$

which is Ito's lemma for $f(z) = z^2$. This gives us a formula for $d(z^2)$. Now we can proceed and choose $u = z$ and $v = z^2$. From our rectangle and the formula for $d(z^2)$ we have

$$
\begin{aligned}
d(z^3) = d(z \cdot z^2) &= zd(z^2) + z^2dz + dzd(z^2) & (2.38) \\
&= z(2zdz + dt) + z^2dz + dz(2zdz + dt) & (2.39) \\
&= 3z^2dz + 3zdt + o(dt) & (2.40)
\end{aligned}
$$

where in the last step we have swept higher order terms into $o(dt)$ and replaced dz^2 by dt. This is Ito's lemma for $f(z) = z^3$.

Continuing in this manner, we can easily derive Ito's lemma for any polynomial of any order in $z(t)$! At that point, we are not a far cry from Ito's lemma for $\mathcal{C}^2$ functions, as they can be approximated by limits of polynomials. Hence, you see that Ito's lemma is actually quite natural if you just remember the rectangle. Furthermore, now you should have an easy time remembering Ito's product rule, $d(uv) = udv + vdu + dudv$.

2.6 SUMMARY

Ito's lemma is the most important result in stochastic calculus for derivative pricing. There are different versions for Brownian motion and for Poisson processes. You should be familiar with both. Roughly speaking, for Brownian motion, since the standard deviation of dz is of order $\sqrt{dt}$, second order terms in dz are of order dt and cannot be ignored. This leads to an extra term in Ito's lemma compared to the ordinary chain rule of calculus. For Poisson processes, most of the time no jumps are occurring and ordinary calculus is fine. However, when a jump occurs, it causes a corresponding jump in any function of the Poisson process. Thus Ito's lemma for Poisson processes is simply a combination of the chain rule of ordinary calculus plus the recognition that when the Poisson process jumps, any function of it has a corresponding jump.

EXERCISES

2.1 Ito's Lemma Practice with Brownian Motion.

Let $dx = xdt + \sqrt{x}dz$. Use Ito's lemma to derive df where $f(x,t)$ is given by:

(a) $f(x,t) = x^2 + t$.

(b) $f(x,t) = \ln(x+t)$.

(c) $f(x,t) = te^x$.

2.2 Ito's Lemma Practice with Poisson Processes.

Let $dx = x^- dt + (x^- - 1)d\pi(\alpha)$ where α is the intensity of the Poisson process. Use Ito's lemma to derive df where $f(x,t)$ is given by:

(a) $f(x,t) = x^2$.

(b) $f(x,t) = t^2 \sqrt{x}$.

(c) $f(x,t) = e^{x^2}$.

2.3 Let $dx = a(b-x)dt + cdz$ and consider $y = e^x$. Use Ito's lemma to find dy and write the stochastic differential equation for y completely in terms of y.

2.4 Consider the following system of Brownian motion driven stochastic differential equations,

$$
\begin{aligned}
dx &= (x+y)dt + xdz_1, \\
dy &= ydt + xdz_2,
\end{aligned}
$$

where $\mathbb{E}[dz_1 dz_2] = \rho dt$. Use Ito's lemma to derive df for a generic twice continuously differentiable $f(x,y,t)$.

2.5 Provide a heuristic derivation (at the level of that given for the Poisson version of Ito's lemma) for the version of Ito's lemma for compound Poisson processes given in equation (2.24).

2.6 Consider the stochastic differential equation

$$dx = \mu x^- dt + \sigma x^- dz + (e^Y - 1)x^- d\pi \qquad (P2.1)$$

where z is Brownian motion, π is a Poisson process with intensity α, and Y is a random variable. Use Ito's lemma to find $d\ln x$.

2.7 Using Ito's Lemma for Vector Processes.

Consider the vector stochastic differential equation

$$dx = a(x)dt + B(x)dz$$

with $x \in \mathbb{R}^2$, and

$$x = \begin{bmatrix} x_1 \\ x_2 \end{bmatrix}, \quad a(x) = \begin{bmatrix} -x_2 \\ x_1 x_2 \end{bmatrix}, \quad B(x) = \begin{bmatrix} x_2 & 0 \\ x_1 & x_2 \end{bmatrix}.$$

Let $z \in \mathbb{R}^2$ with

$$z = \begin{bmatrix} z_1 \\ z_2 \end{bmatrix}$$

be a vector Brownian motion with covariance structure $\mathbb{E}[dz dz^T] = \Sigma dt$ where

$$\Sigma = \begin{bmatrix} 0.2 & 0.01 \\ 0.01 & 0.4 \end{bmatrix}.$$

Take $f(x,t) = x_1^2 + 2x_1 x_2$. In this problem, we will apply the vector Ito formula in (2.31).

(a) Calculate the gradient and Hessian matrix of f. That is, compute f_x and f_{xx} as given in (2.32).

(b) Use the results from part (a) and apply the Ito's lemma for vector processes formula in (2.31) to calculate df.

2.8 Let

$$dx = adt + bdz_1 \qquad (P2.2)$$

and

$$dy = fdt + gdz_2 \qquad (P2.3)$$

where z_1 and z_2 are correlated Brownian motions with correlation coefficient ρ. (i.e. $\mathbb{E}(dz_1 dz_2) = \rho dt$.)

(a) Use Ito's lemma to find $d(xy)$.

(b) Use Ito's lemma to find $d(x/y)$.

2.9 Compute

$$\int_0^T z(t)dz(t) \tag{P2.4}$$

where $z(t)$ is Brownian motion. (Hint: consider Ito's lemma for z^2.)

2.10 Counter-Intuition.

Let

$$dx = axdt + bxdz. \tag{P2.5}$$

Assume that $a > 0$. That is, the growth rate of $\mathbb{E}[x(t)]$ is positive. Now consider $1/x(t)$. Is it possible for $\mathbb{E}[1/x(t)]$ to also have a positive growth rate? Give conditions on a and b for when this is possible. Intuitively, provide an explanation for this.

2.11 Intuition behind Ito's Lemma for Brownian Motion.

(This intuitive look at Ito's lemma was communicated to me by Muruhan Rathinam, circa 2005.) Let

$$dx = adt + bdz, \tag{P2.6}$$

then from Ito's lemma $f(x,t)$ satisfies

$$df = (f_t + af_x + \frac{1}{2}b^2 f_{xx})dt + bf_x dz. \tag{P2.7}$$

Recall our heuristic derivation based on a Taylor expansion,

$$df = f_t dt + f_x dx + \frac{1}{2}f_{xx}(dx)^2 + \dots \tag{P2.8}$$

or

$$df = f_t dt + f_x(adt + bdz) + \frac{1}{2}f_{xx}(a^2(dt)^2 + b^2(dz)^2 + 2abdtdz) + \dots \tag{P2.9}$$

(a) Compute the mean of df to lowest order in dt using the terms of the Taylor expansion shown above.

(b) Compute the standard deviation of the terms that contain randomness in the above expansion. That is, compute the standard deviation of the dz, $dtdz$, and $(dz)^2$ terms separately. Which term has standard deviation of lowest order in dt? (Hint: The fourth moment of a $\mathcal{N}(0, \sigma^2)$ is $3\sigma^4$.)

(c) Use this analysis to argue for the plausibility of Ito's lemma in (P2.7).

2.12 Show that (2.29) follows from (2.31).

CHAPTER 3

Stochastic Differential Equations with Solutions

I N this chapter, we review some stochastic differential equations that have closed form solutions. These are also some of the stochastic differential equations used for modeling asset prices and other relevant financial variables. In these solutions, note the role that Ito's lemma plays. Most importantly, not many stochastic differential equations have closed form solutions. Thus, these are stochastic differential equations that everyone should know!

3.1 GEOMETRIC BROWNIAN MOTION

Geometric Brownian motion is represented by the following stochastic differential equation where a and b are constants,

$$dx = axdt + bxdz. \tag{3.1}$$

A closed form solution to geometric Brownian motion can be found as follows. By Ito's lemma with $f(x) = \ln(x)$ we have

$$d\ln(x) = (a - \frac{1}{2}b^2)dt + bdz. \tag{3.2}$$

Since a and b are constants, integrating from 0 to t gives

$$\ln(x(t)) - \ln(x(0)) = (a - \frac{1}{2}b^2)t + bz(t), \tag{3.3}$$

and finally solving for $x(t)$ by exponentiation results in

$$x(t) = e^{(a-\frac{1}{2}b^2)t+bz(t)}x(0). \tag{3.4}$$

Note that geometric Brownian motion is a lognormal process. That is, from (3.3) in log-coordinates $x(t)$ is a Gaussian process with drift rate $a - \frac{1}{2}b^2$ and variance rate b^2. Figure 3.1 shows a typical sample path of geometric Brownian motion.

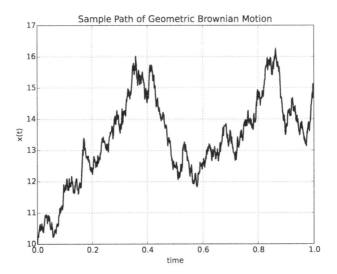

Figure 3.1 Typical sample path of geometric Brownian motion.

3.1.1 Stock Price Interpretation

Geometric Brownian motion is the "standard model" for *continuous* stock price movements. It comes from the following interpretation. Let $x(t)$ be the price of a stock. Then, over the time period dt, the instantaneous return r on the stock is given by

$$r = \frac{x(t+dt) - x(t)}{x(t)} = \frac{dx}{x}. \tag{3.5}$$

Geometric Brownian motion models this instantaneous return as

$$\frac{dx}{x} = adt + bdz \tag{3.6}$$

where a and b are constants. That is, the instantaneous return r over the next dt is Gaussian with

$$\mathbb{E}[r] = adt, \quad Var(r) = b^2 dt. \tag{3.7}$$

Thus, this is a very natural model of asset prices.

Using this interpretation, the parameter a corresponds to the annualized expected rate of return of the stock, while b^2 is the annualized variance of the return. The annualized standard deviation of the return is simply b, and is commonly referred to as the *volatility* of the stock.

Example 3.1 (Solving Geometric Brownian Motion)
Consider a stock with price $S(t)$. Assume that we model the movement of this stock price with the stochastic differential equation,

$$dS = 0.15S(t)dt + 0.3S(t)dz.$$

From our model, this stock has an annualized expected rate of return of 15% and an annualized variance of return of $0.3^2 = 0.09$. The annualized standard deviation of the return (i.e., the *volatility*) of the stock is 30%. Additionally, the closed form solution for the stock price is

$$S(t) = S(0)e^{\left(0.15 - \frac{1}{2}(0.3)^2\right)t + 0.3z(t)} = S(0)e^{0.105t + 0.3z(t)}.$$

3.2 GEOMETRIC POISSON MOTION

Geometric Poisson motion is the equivalent of geometric Brownian motion, but driven by a Poisson process,

$$dx = ax^- dt + (b-1)x^- d\pi. \tag{3.8}$$

The reason for the term $b - 1$ is as follows. If the current value is x^- and a jump occurs, we will jump by an amount $dx = (b-1)x^-$. Thus, we jump from x^- to $x^- + dx = x^- + (b-1)x^- = bx^-$. Hence, by writing $(b-1)$ we think of b as indicating the multiple of the current state that results if a jump occurs. That is, a jump leads to the transition $x^- \to bx^-$. Note that if we didn't use this convention, things would be a bit messier.

Again, there exists a closed form solution that is obtained by changing to log coordinates. By Ito's lemma with $f(x) = \ln(x)$, we have

$$d\ln(x) = a\,dt + \ln(b)d\pi. \tag{3.9}$$

Integrating both sides leads to

$$\ln(x(t)) - \ln(x(0)) = at + \ln(b)\pi(t). \tag{3.10}$$

This is a Poisson process plus drift in log coordinates. If we solve for $x(t)$, we obtain

$$x(t) = e^{at + \ln(b)\pi(t)}x(0) = b^{\pi(t)}e^{at}x(0). \tag{3.11}$$

From this solution, we can see that $x(t)$ drifts at an exponential rate of a, but gets multiplied by b every time a jump occurs. The quantity $\pi(t)$ just counts the number of jumps that happen from time 0 to t.

3.2.1 Conditional Lognormal Version

The following stochastic differential equation uses a compound Poisson process with a lognormal jump size. This results in a *conditional* lognormal process that is different from geometric Brownian motion. It is modeled as

$$dx = ax^- dt + (Y-1)x^- d\pi, \tag{3.12}$$

where $Y = e^Z$ is a lognormal random variable with Z normally distributed with mean ν and variance σ^2.

Once again, a change to log coordinates facilitates finding a closed form solution. Using Ito's lemma with $f(x) = \ln(x)$ gives

$$d \ln x = adt + \ln(Y)d\pi. \tag{3.13}$$

Integrating both sides yields

$$\ln x(t) - \ln x(0) = at + \int_0^t \ln(Y)d\pi = at + \sum_{i=1}^{\pi(t)} \ln(Y_i). \tag{3.14}$$

Finally, exponentiation and rearrangement result in

$$x(t) = \left(e^{at} \prod_{i=1}^{\pi(t)} Y_i \right) x(0). \tag{3.15}$$

But since $Y = e^Z$ is lognormal, we can write $\prod_{i=1}^{\pi_t} Y_i = e^{\sum_{i=1}^{\pi(t)} Z_i}$ where each Z_i is normal. Hence we have

$$x(t) = \left(e^{at + \sum_{i=1}^{\pi(t)} Z_i} \right) x(0) \tag{3.16}$$

which, conditioned on the value of $\pi(t)$, is lognormally distributed.

3.3 JUMP DIFFUSION MODEL

The following stochastic differential equation uses a compound Poisson process with a lognormal jump size, as given above, and also drives it with a Brownian motion. This results in a process that involves jumps and a diffusion (Brownian) term, and is still conditionally lognormal. It is known as a jump diffusion model. The stochastic differential equation describing this model is given by

$$dx = ax^- dt + bx^- dz + (Y - 1)x^- d\pi, \tag{3.17}$$

where Y is a lognormal random variable.

Using Ito's lemma with $f(x) = \ln(x)$ allows one to write the solution in closed form as

$$x(t) = \left(e^{(a - \frac{1}{2}b^2)t + bz(t)} \right) \left(\prod_{i=1}^{\pi(t)} Y_i \right) x(0). \tag{3.18}$$

But since Y is lognormal, each jump Y_i can be written as $Y_i = e^{Z_i}$ where Z_i is a normal random variable. This allows us to write the product of Y_i's as

$$\prod_{i=1}^{\pi(t)} Y_i = \prod_{i=1}^{\pi(t)} e^{Z_i} = e^{\sum_{i=1}^{\pi(t)} Z_i}.$$

Thus, the closed form solution can also be written as

$$x(t) = \left(e^{(a - \frac{1}{2}b^2)t + bz(t) + \sum_{i=1}^{\pi(t)} Z_i} \right) x(0) \tag{3.19}$$

which, conditioned on the number of jumps, $\pi(t)$, follows the lognormal distribution.

This model is nice because it can produce distributions that have heavier tails (i.e., more extreme price movements) than the lognormal distribution which is commonly observed in empirical data.

3.3.1 Jump Diffusion as a Stock Price Model

When used as the model for a stock price $S(t)$, the jump diffusion process leads to the stock's instantaneous return being

$$r = \frac{dS}{S^-} = adt + bdz + (Y-1)d\pi,$$

which is the sum of three parts. The term adt is a deterministic drift, bdz is a Gaussian term, and finally $(Y-1)d\pi$ allows random jumps to occur which could represent sudden unexpected events such as earnings surprises or the like.

3.4 A MORE GENERAL STOCHASTIC DIFFERENTIAL EQUATION

A slightly more general description of a stochastic differential equation driven only by Brownian motion is

$$dx = (a+bx)dt + (c+fx)dz, \tag{3.20}$$

where a, b, c, and f are constants. We will obtain the solution to this stochastic differential equation using the concept of an integrating factor that arises in ordinary differential equations [6]. First, write the stochastic differential equation as

$$dx - bxdt - fxdz = adt + cdz. \tag{3.21}$$

In ordinary differential equations, we would try to make the left side an exact differential by using an integrating factor. The integrating factor is usually related to the solution to the differential equation if the right side of (3.21) is set to zero (the homogeneous solution). That is,

$$dx - bxdt - fxdz = 0.$$

This is just geometric Brownian motion, so the solution is

$$x(t) = x(0)e^{(b-\frac{1}{2}f^2)t + fz(t)}.$$

Hence, mimicking the approach in ordinary differential equations, we will try an integrating factor of the form

$$e^{-(b-\frac{1}{2}f^2)t - fz(t)}.$$

Therefore, let us compute the total differential of the integrating factor multiplied by $x(t)$,

$$
\begin{aligned}
d(e^{-(b-\frac{1}{2}f^2)t-fz(t)}x) &= e^{-(b-\frac{1}{2}f^2)t-fz(t)}\left(-(b-\frac{1}{2}f^2)xdt - fxdz + \frac{1}{2}f^2xdt\right.\\
&\qquad\qquad\qquad\left. +(a+bx)dt + (c+fx)dz - f(c+fx)\,dt\right)\\
&= e^{-(b-\frac{1}{2}f^2)t-fz(t)}\left((a-fc)dt + cdz\right).
\end{aligned}
$$

Integrating both sides from 0 to t leads to

$$
e^{(-(b-\frac{1}{2}f^2)t-fz(t))}x(t) - x(0) = \int_0^t e^{(-(b-\frac{1}{2}f^2)s-fz(s))}((a-fc)ds + cdz(s))
$$

and rearrangement gives the final form

$$
x(t) = e^{((b-\frac{1}{2}f^2)t+fz(t))}x(0) + \int_0^t e^{((b-\frac{1}{2}f^2)(t-s)+f(z(t)-z(s)))}((a-fc)ds + cdz(s)).
$$

This result may appear a bit messy, but it is handy to know that the solution to stochastic differential equations of this form can be represented this way.

3.4.1 Ornstein–Uhlenbeck Process and Mean Reversion

A special case of the above stochastic differential equation is the so-called Ornstein–Uhlenbeck process [50]. This is a Gaussian model in which a price or financial variable is mean reverting. That is, the process is pulled toward a specified level. It is given by

$$
dx = -b(x-a)dt + cdz \tag{3.22}
$$

where a is the level toward which $x(t)$ reverts, and b is the mean reversion rate. Thus, $x(t)$ is drawn toward a at a rate of b. This is easily seen by analyzing the drift term $-b(x-a)$. When $x > a$, the drift term is negative and the process is moving down on average, while when $x < a$ the drift term is positive and the process is moving up on average. Only at $x = a$ is the drift zero. The Brownian driven term cdz just adds Gaussian noise.

A sample path of the mean reverting Ornstein–Uhlenbeck process is depicted in Figure 3.2 with the parameter values $a = 2$, $b = 5$, $c = 0.5$ and initial condition $x(0) = 3$. Note how the process is drawn toward the level $a = 2$.

In a finance context, mean reverting processes of this specific form are sometimes also called Vasicek models. Vasicek used this type of process to capture the movement of the short rate of interest in term structure modeling [51]. In that context, we will encounter this model in Chapter 7, Section 7.3.2.1.

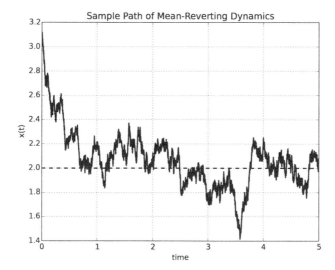

Figure 3.2   Simulation of mean-reverting dynamics, $a = 2$, $b = 5$, $c = 0.5$, $x(0) = 3$.

Since this mean reverting process is of the form of (3.20), it has a closed form solution given as

$$
\begin{aligned}
x(t) &= e^{-bt}x(0) + \int_0^t e^{-b(t-s)}(abds + cdz(s)) \\
&= e^{-bt}x(0) + a\left(1 - e^{-bt}\right) + \int_0^t e^{-b(t-s)}cdz(s).
\end{aligned}
$$

Observe that $x(t)$ is a Gaussian process as well. Roughly speaking, this is because the integral involving $dz(s)$ has a non-random integrand, and overall the integral can be thought of as a "sum of Gaussians," which is Gaussian. You are encouraged to work through this logic in one of the chapter exercises.

One disadvantage of this process is that values of $x(t)$ can become negative because the Gaussian distribution always has some probability of being negative. This can be undesirable because prices and quantities such as interest rates should generally not be negative.

3.5   COX–INGERSOLL–ROSS PROCESS

Another version of a mean reverting process is the Cox–Ingersoll–Ross (CIR) process [12], given by

$$dx = -b(x - a)dt + c\sqrt{x}dz \tag{3.23}$$

and used often in short rate models or stochastic volatility type models. Note that it is very similar to the Ornstein–Uhlenbeck process, except that the driving Brownian motion is multiplied by $\sqrt{x}$. This makes the noise dependent on the size of x. Furthermore, with correct parameter values, this process will always be positive.

The CIR process is actually related to the Ornstein–Uhlenbeck process as follows. Consider n Ornstein–Uhlenbeck processes

$$dy_1 = -\frac{1}{2}\alpha y_1 dt + \frac{1}{2}\beta dz_1 \tag{3.24}$$

$$\vdots \tag{3.25}$$

$$dy_n = -\frac{1}{2}\alpha y_n dt + \frac{1}{2}\beta dz_n \tag{3.26}$$

where the Brownian motions $z_1, \ldots, z_n$ are uncorrelated. Now, consider the process

$$x(t) = y_1^2(t) + \ldots + y_n^2(t). \tag{3.27}$$

By the multivariable Ito's lemma, $x(t)$ follows

$$
\begin{aligned}
dx &= \sum_{i=1}^{n} 2y_i(t)\left(-\frac{1}{2}\alpha y_i dt + \frac{1}{2}\beta dz_i\right) + \sum_{i=1}^{n}\left(\frac{\beta^2}{4}\right) dt \\
&= \sum_{i=1}^{n}\left(-\alpha y_i^2 + \frac{\beta^2}{4}\right) dt + \sum_{i=1}^{n}(\beta y_i dz_i) \\
&= \left(-\alpha x(t) + n\frac{\beta^2}{4}\right) dt + \beta \sum_{i=1}^{n}(y_i dz_i).
\end{aligned}
$$

We can perform a little trick by writing the last term as

$$\beta\sqrt{x(t)}\sum_{i=1}^{n}\left(\frac{y_i}{\sqrt{x(t)}}dz_i\right).$$

Surprisingly, it turns out that

$$\sum_{i=1}^{n}\left(\frac{y_i}{\sqrt{x(t)}}dz_i\right) \tag{3.28}$$

is actually a Brownian motion!

How can we see this? Well, note that if we interpret each Brownian increment as being normally distributed with mean zero and variance dt, i.e. $dz_i \sim \mathcal{N}(0, dt)$, then (3.28) is just the sum of Gaussians. But, it is a well known property that the sum of Gaussians is Gaussian. Thus, (3.28) is normally distributed. For it to be the increment of Brownian motion, heuristically at least, we only need to show that the mean is zero and the variance is dt. To compute the mean we have

$$\mathbb{E}\left[\sum_{i=1}^{n}\left(\frac{y_i}{\sqrt{x(t)}}dz_i\right)\right] = \sum_{i=1}^{n}\left(\frac{y_i}{\sqrt{x(t)}}\mathbb{E}[dz_i]\right) = 0, \tag{3.29}$$

and the variance is computed as

$$Var\left[\sum_{i=1}^{n}\left(\frac{y_i}{\sqrt{x(t)}}dz_i\right)\right]=\sum_{i=1}^{n}\left(\frac{y_i^2}{x(t)}Var(dz_i)\right)=\sum_{i=1}^{n}\left(\frac{y_i^2}{x(t)}dt\right)=\frac{dt}{x(t)}\sum_{i=1}^{n}y_i^2=dt$$

(3.30)

where we used equation (3.27) and the fact that the dz_i's are independent. Thus we can actually use the replacement

$$dz=\sum_{i=1}^{n}\left(\frac{y_i}{\sqrt{x(t)}}dz_i\right)$$

(3.31)

to convert (3.28) to

$$dx=\left(-\alpha x(t)+n\frac{\beta^2}{4}\right)dt+\beta\sqrt{x(t)}dz.$$

(3.32)

Now, it turns out that for $n\geq 2$, we are guaranteed to have a positive process (otherwise the process can and will reach zero at times!). Hence, we should always make sure that our parameter choices correspond to selecting $n\geq 2$.

This is easily done since we can match coefficients in (3.23) and (3.32). This gives

$$ab=\frac{n\beta^2}{4},\quad b=\alpha,\quad c=\beta.$$

(3.33)

From these relationships, we obtain $n=\frac{4ab}{c^2}$.

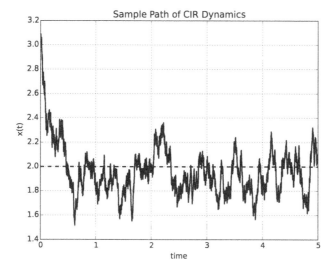

Figure 3.3 Simulation of CIR dynamics, $a=2$, $b=5$, $c=0.5$, $x(0)=3$.

For n corresponding to an integer, we have the solution as the sum of the square of Ornstein–Uhlenbeck processes, which are Gaussian distributed. Thus, overall, the process $x(t)$ will be chi-squared distributed [15]. A sample path of a CIR process is depicted in Figure 3.3 for the parameter values $a - 2$, $b - 5$, $c - 0.5$, and initial condition $x(0) = 3$.

3.6   SUMMARY

In this chapter we explored some of the standard stochastic differential equation models used in finance. It is important to have a feel for these processes and how they are used to model phenomena in finance. Moreover, note that most of them correspond to some transformation of Brownian motion or the Poisson process. That is, you can think of these models as building more and more complicated models from our basic building blocks of Brownian motion and the Poisson process.

EXERCISES

3.1 Modeling a Bond.

(a) Let r_0 be an interest rate quoted yearly, and let time be measured in years. Moreover, let $r_0 \Delta t$ be the interest applied over the time period Δt. Write the equation for the return of a bond B over the time period t to $t + \Delta t$, where the bond earns the yearly rate r_0.

(b) Argue that the limit of your equation in (a) is the ordinary differential equation $\frac{dB}{dt} = r_0 B$.

(c) Assume that you place the amount $B(0)$ in a risk-free bond at time 0. Solve the resulting differential equation in (b) to obtain the value in the bond $B(t)$ at time t.

3.2 Expected Return.

Consider a stock whose price follows geometric Brownian motion,

$$dS = \mu S dt + \sigma S dz.$$

Consider representing the solution as

$$S(0)e^{rt} = S(t)$$

where we can think of r as the annualized random return of the stock. We can then compute the continuously compounded *expected annualized return* as

$$\nu = \mathbb{E}\left[\frac{1}{t}\ln\left(\frac{S(t)}{S(0)}\right)\right].$$

(a) What is the continuously compounded expected annualized return under geometric Brownian motion?

(b) Compute the continuously compounded expected annualized return for $\mu = 0.2$ and $\sigma = 0,\ \ 0.1,\ \ 0.3,$ and 0.5. What happens to this expected return as the volatility increases?

3.3 Lognormal Random Variables.

Let $Z \sim \mathcal{N}(\mu, \sigma^2)$ and show that $\mathbb{E}[e^Z] = e^{\mu + \frac{1}{2}\sigma^2}$.

3.4 Moments of the Lognormal Random Variable.

Use the results of Exercise 3.3 to show that the k-th moment of the lognormal random variable $Y = e^Z$ with $Z \sim \mathcal{N}(\mu, \sigma^2)$ is $\mathbb{E}[Y^k] = e^{k\mu + \frac{1}{2}k^2\sigma^2}$.

3.5 Let $S(t)$ follow geometric Brownian motion,

$$dS = \mu S dt + \sigma S dz.$$

Assume that $S(0) = S_0$. Use Exercise 3.4 to compute the following:

(a) $\mathbb{E}[S(t)]$.

(b) $\mathbb{E}[S^2(t)]$.

(c) $Var(S(t))$.

3.6 Write the form of the solution to the stochastic differential equation, $dx = \mu dt + \sigma x dz$.

3.7 Consider $x(t) = \int_0^t f(s) dz(s)$.

(a) Argue that $x(t)$ should be Gaussian distributed. (Hint: Approximate the integral as a sum and use that fact that the sum of Gaussian random variables is Gaussian.)

(b) Compute $\mathbb{E}[x(t)]$ and $Var(x(t))$.

3.8 Consider the jump diffusion model

$$\frac{dS}{S^-} = a dt + b dz + (Y - 1) d\pi(\alpha)$$

where α is the intensity of the Poisson process, and $Y = e^Z$ with $Z \sim \mathcal{N}(\nu_J, \sigma_J^2)$. By conditioning on the number of jumps that have occurred by time t, show that the expectation of $x(t)$ can be expressed as

$$\mathbb{E}[x(t)] = \sum_{i=0}^{\infty} \mathbb{E}[x(t)|\pi(t) = k]\mathbb{P}(\pi(k) = k) = \sum_{k=0}^{\infty} \frac{\alpha^k e^{-\alpha}}{k!} e^{\nu t + k\nu_J + \frac{1}{2}(b^2 t + k\sigma_J^2)}.$$

3.9 Use the solution of the Ornstein–Uhlenbeck process in equation (3.23) to compute its expectation, $\mathbb{E}[x(t)]$. (Hint: Use the fact that the expectation of the integral term $\int (\cdot) dz$ is zero.) What is $\lim_{t \to \infty} \mathbb{E}[x(t)]$?

3.10 Find an ordinary differential equation for the expectation of the Ornstein–Uhlenbeck process $\mathbb{E}[x(t)]$ as follows. Consider the Ornstein–Uhlenbeck stochastic differential equation,

$$dx = -b(x - a)dt + cdz,$$

and take the expectation of both sides of this equation. Use the fact that $\mathbb{E}[dz] = 0$ and switch the "d" and expectation on the left side to obtain an ordinary differential equation for $\mathbb{E}[x(t)]$.

(a) What ordinary differential equation do you obtain?

(b) Solve the differential equation in (a). Does it correspond to your answer in Exercise 3.9?

(c) Compute the steady state value of the expectation by setting $\frac{d\mathbb{E}[x(t)]}{dt} = 0$.

CHAPTER 4

The Factor Model Approach to Arbitrage Pricing

IN this chapter, we present absence of arbitrage conditions when returns are described by linear factor models. In doing so, we rely heavily on Ross' Arbitrage Pricing Theory [46], but tailored to the context of derivative pricing. What we find is that by interpreting stochastic differential equations as instantaneous factor models, a simple factor model based absence of arbitrage condition is applicable. Moreover, this condition provides a direct and transparent route to obtaining the absence of arbitrage equations that govern the pricing of a vast array of derivatives. To back up this claim, in subsequent chapters we will derive these equations in a great number of situations. But first we must understand the underlying principle, which is in fact quite simple.

4.1 RETURNS AND FACTOR MODELS

Most of the modeling in asset pricing theory is done using *linear factor models*. We adopt that same paradigm here. That is, a return on an asset is assumed to be of the form

$$r = \alpha + \beta f \qquad (4.1)$$

where α and β are constants and f is a random variable called a *factor*. We have written a single factor model since there is only one f. However, we can just as easily consider multifactor models of the form

$$r = \alpha + \sum_{i-1}^{n} \beta_i f_i \qquad (4.2)$$

where each f_i, $i = 1, ..., n$ is a random factor.

4.1.1 Returns from Price Changes

Above, I said we would model returns as factor models. Where will these models come from? Well, a return is typically calculated from the change in price of an asset. Given a period of time Δt, the return on an asset is defined as

$$r = \frac{P(t + \Delta t) - P(t)}{P(t)}, \tag{4.3}$$

where $P(t)$ is the dollar amount invested at time t and $P(t + \Delta t)$ is the new dollar value at time $t + \Delta t$. Figure 4.1 shows the cash flow diagram corresponding to this, where the time increment is denoted as dt instead of Δt.

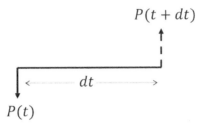

Figure 4.1   Typical cash flow diagram for an asset.

Taking the hint from Figure 4.1, if we consider the return over smaller and smaller time increments, we can think of the "instantaneous" return on P as

$$r = \frac{P(t + dt) - P(t)}{P(t)} = \frac{dP(t)}{P(t)}$$

where dP will come from a stochastic differential equation model.

4.1.2   Stochastic Differential Equations and Factor Models

As you know, in finance we often like to model asset prices as stochastic differential equations. For instance, a stock price could be assumed to follow geometric Brownian motion,

$$dS = \mu S dt + \sigma S dz, \tag{4.4}$$

as explained in the previous chapter. This is a model of a price change over time dt. Hence, I should be able to use it to compute the instantaneous return of an asset over time dt as

$$r = \frac{S(t + dt) - S(t)}{S(t)} = \frac{dS(t)}{S(t)} = \mu dt + \sigma dz. \tag{4.5}$$

This is also an example of a factor model! In the above formula, if we associate $\alpha = \mu dt$, $\beta = \sigma$, and $f = dz$ then it is even a standard single-factor linear factor model!

This idea of viewing stochastic differential equations for prices as a route to factor models for returns can be pursued more generally. If I am given a stochastic differential equation,

$$dx(t) = a(x(t), t)dt + b(x(t), t)dz \tag{4.6}$$

and it is describing price changes over time period dt, I can use it to determine a model of returns,

$$r = \frac{dx(t)}{x(t)} = \frac{a(x(t),t)}{x(t)}dt + \frac{b(x(t),t)}{x(t)}dz. \qquad (4.7)$$

This also looks like a factor model with the factor being dz. In fact, I like to write it in the form,

$$r = \alpha dt + \beta dz \qquad (4.8)$$

where $\alpha = \frac{a(x(t),t)}{x(t)}$ and $\beta = \frac{b(x(t),t)}{x(t)}$ (and arguments are being suppressed). This is slightly different from the standard factor model in that I explicitly separate out dt in the first term. I like to go even a little further and interpret dt as a special factor which is non-random. This is purely for convenience, but I will use this convention throughout.

Another note is that α and β don't seem to be constant! However, we should view a factor model from a stochastic differential equation as being valid only over dt and conditioned upon information at time t, including $x(t)$ (or t^- and $x(t^-)$ if appropriate). Therefore, conditioned upon information at time t, $x(t)$ is known and α and β are then known constants over the time period t to $t + dt$.

The take-away is that stochastic differential equations lead to instantaneous factor models that apply over time increments of size dt. In what follows, I will use notation that is indicative of stochastic differential equations in that factors will be denoted by dz and I will include a special time factor dt when I describe returns.

4.2 FACTOR APPROACH TO ARBITRAGE USING RETURNS

To derive absence of arbitrage conditions, we will consider the returns of *tradable* assets. By a tradable asset, I mean an asset that you can actually buy. For example, you can buy a bond or a share of stock. If the stock that you buy also pays a dividend, then by purchasing a share you have purchased the dividend stream as well. In such a case, note that you cannot just purchase the "price" of the stock. You are stuck with everything that comes along with the share (such as dividends).

Consider the returns of n tradable assets, denoted r_i, $i = 1 \ldots n$, and assume that each return follows a multifactor model driven by m random factors dz_j, $j = 1 \ldots m$. That is, each return r_i can be modeled as

$$r_i = \alpha_i dt + \sum_{j=1}^{m} \beta_{ij} dz_j.$$

To make the notation more compact, we will utilize vector/matrix representations with

$$r = \begin{bmatrix} r_1 \\ r_2 \\ \vdots \\ r_n \end{bmatrix}, \quad \alpha = \begin{bmatrix} \alpha_1 \\ \alpha_2 \\ \vdots \\ \alpha_n \end{bmatrix}, \quad \beta = \begin{bmatrix} \beta_{11} & \cdots & \beta_{1m} \\ \beta_{21} & \cdots & \beta_{2m} \\ \vdots & \ddots & \vdots \\ \beta_{n1} & \cdots & \beta_{nm} \end{bmatrix}, \quad dz = \begin{bmatrix} dz_1 \\ dz_2 \\ \vdots \\ dz_m \end{bmatrix}. \quad (4.9)$$

Under this notation, the factor models of the n tradable assets can be succinctly written as

$$r = \alpha dt + \beta dz \quad (4.10)$$

where, as in (4.9), $r \in \mathbb{R}^n$, $\alpha \in \mathbb{R}^n$, $\beta \in \mathbb{R}^{n \times m}$, and $dz \in \mathbb{R}^m$.

Now consider a portfolio of these assets. We will represent the portfolio by a vector $x \in \mathbb{R}^n$,

$$x = \begin{bmatrix} x_1 \\ x_2 \\ \vdots \\ x_n \end{bmatrix}$$

where each element x_i denotes the *dollar amount* invested in tradable asset i. Hence, the total cost of purchasing this portfolio is just the sum of the x_i's, which can be written as

$$\text{cost} = \sum_{i=1}^{n} x_i = x^T \mathbf{1} \quad (4.11)$$

where $\mathbf{1} \in \mathbb{R}^n$ is a vector with every element equal to 1.

Over the time period dt, the returns in (4.10) indicate the percentage change in value of each tradable asset. Hence, the total profit/loss in our portfolio over the time period dt is given by

$$\text{profit/loss} = \sum_{i=1}^{n} x_i r_i = x^T r. \quad (4.12)$$

Note that in this equation, there is no requirement that the dollar amount purchased, x_i, be positive. When $x_i < 0$, we are "short" asset i in an amount equal to x_i. To obtain a better feel for this equation, consider the following example.

Example 4.1 (Calculating Profit/Loss)
Consider a portfolio with an amount $x_1 = \$5$ in the first asset and $x_2 = \$8$ in the second asset. If the returns on the first and second asset are $r_1 = 3\%$ and $r_2 = -5\%$, respectively, then I make $x_1 r_1 = (\$5)(3\%) = \$0.15$ on asset 1 and lose $x_2 r_2 = (\$8)(-5\%) = \$0.40$ on asset 2. Overall, my profit/loss is

$$\text{profit/loss} = x_1 r_1 + x_2 r_2 = 0.15 - 0.40 = -\$0.25.$$

The dollar amount invested can also be negative, and the profit/loss equation still holds. This time consider a portfolio that is "short" asset 1 in an amount $x_1 = -\$4$

and "long" asset 2 in an amount $x_2 = \$7$. If the returns on asset 1 and asset 2 are still $r_1 = 3\%$ and $r_2 = -5\%$, then the profit/loss on asset 1 is $x_1 r_1 = (-\$4)(3\%) = -\$0.12$ and on asset 2 is $x_2 r_2 = (\$7)(-5\%) = -\$0.35$. The overall profit/loss is just the sum of the individual profits and losses,

$$\text{profit/loss} = x_1 r_1 + x_2 r_2 = -0.12 - 0.35 = -\$0.47.$$

4.2.1 Arbitrage

We can analyze the components of the profit/loss of a portfolio in a little more detail by writing out r in terms of its factor model,

$$x^T r = x^T(\alpha dt + \beta dz) = (x^T \alpha)dt + (x^T \beta)dz. \tag{4.13}$$

By doing this, we see that there is a component of the return of the portfolio that is risk-free, $(x^T \alpha)dt$, and a component that is driven by the risky factors, $(x^T \beta)dz$. Moreover, we can now see that the overall profit/loss will actually be riskless as long as

$$x^T \beta = 0 \tag{4.14}$$

since that is the coefficient of the dz term which is injecting the randomness. If $x^T \beta = 0$ then the profit/loss on this portfolio is deterministic and given by $x^T \alpha dt$.

With this idea in mind, let us consider a simple portfolio. In particular, I will choose this portfolio so that it costs nothing and has no risk. That is

$$x^T \mathbf{1} \;=\; 0 \quad \text{No cost.} \tag{4.15}$$
$$x^T \beta \;=\; 0 \quad \text{No risk.} \tag{4.16}$$

But, if something costs nothing and has no risk, then it better not make a profit! If it did make a profit, then this is what we would call an *arbitrage* opportunity. I would be *guaranteed* (no risk) to make money (profit) without ever having to use any of my own money (no cost). That would be too good to be true!

To present this idea a little more formally, let's define an arbitrage portfolio.

(★) Arbitrage Portfolio
An arbitrage portfolio is a specification of the dollar amounts invested in each asset $x \in \mathbb{R}^n$ that satisfies the three equations:

$$x^T \mathbf{1} = 0 \quad \textit{No cost.}$$
$$x^T \beta = 0 \quad \textit{No risk.}$$
$$x^T \alpha > 0 \quad \textit{Profit.}$$

The whole basis for derivative pricing is that arbitrage portfolios should not exist. That is, no matter what portfolio x I create, if it has no cost and no risk, it must not be able to make a profit. Note that it can't make a loss either, because then $-x$ would be an arbitrage. Therefore, to have no arbitrage, my portfolio must have no profit or loss. This key condition is stated as the following implication, which, unfortunately, is not all that useful...

(★) A (not very useful) Necessary Absence of Arbitrage Condition

A necessary condition for no arbitrage is for the following implication to be true:

$$\left.\begin{array}{ll} x^T \mathbf{1} = 0 & No\ cost \\ x^T \beta = 0 & No\ risk \end{array}\right\} \Rightarrow x^T \alpha = 0 \quad No\ profit/loss. \tag{4.17}$$

Note that this implication can be written entirely in a compact matrix form as

$$\left[\begin{array}{c} \mathbf{1^T} \\ \beta^T \end{array}\right] x = 0 \Rightarrow \alpha^T x = 0. \tag{4.18}$$

As the name indicates, the above condition for absence of arbitrage is not terribly useful. However, an equivalent condition is extremely useful and will provide the foundation for our derivations of absence of arbitrage equations for derivative securities. In fact, the condition which we will present next is astonishingly useful! But to derive it, we need to first recall some linear algebra relationships.

4.2.2 Null and Range Space Relationship

Note that the condition (4.18) can be written in a linear algebra context as

$$Ax = 0 \Rightarrow \alpha^T x = 0 \text{ where } A = \left[\begin{array}{c} \mathbf{1^T} \\ \beta^T \end{array}\right]. \tag{4.19}$$

Now, the set of all vectors x such that $Ax = 0$ is known as the *null* space of the matrix, $\mathbf{N}(A)$, that is

$$\mathbf{N}(A) = \{x | Ax = 0\}. \tag{4.20}$$

On the other hand, the set of all vectors y such that there exists an x with $y = Ax$ is known as the *range* space of the matrix, $\mathbf{R}(A)$, that is

$$\mathbf{R}(A) = \{y \mid \exists\ x \text{ such that } Ax = y\ \}. \tag{4.21}$$

Finally, we recall the notion of the perpendicular set of a given set of vectors. Let M be a set of vectors, then $M^\perp$ is the set of all vectors z such that z is orthogonal to all vectors in M. For two vectors x and z to be orthogonal, we mean that $x^T z = 0$. Thus, $M^\perp$ is defined as

$$M^\perp = \left\{z \mid z^T x = 0,\ \forall x \in M\right\}. \tag{4.22}$$

In order to derive a useful condition from (4.18), we will use the following relationship between the null and range space of a matrix.

(★) Null and Range Space Relationship: $\mathbf{N}(A)^\perp = \mathbf{R}(A^T)$.

Proof: The proof of this is rather simple.

$$x \in \mathbf{N}(A) \Rightarrow Ax = 0 \Rightarrow y^T A x = 0 \text{ for all } y.$$

Then

$$y^T A x = (A^T y)^T x = 0,$$

which means that $A^T y$ is orthogonal to the null space of A for all y. That is exactly saying that

$$\mathbf{N}(A)^\perp = \mathbf{R}(A^T).$$

□

4.2.3   A Useful Absence of Arbitrage Condition

Using the null and range space relationship, we can convert (4.18) into a very useful condition. Its importance cannot be overstated. Hence it receives two stars!!

(★★) Return APT: A Useful Necessary Absence of Arbitrage Condition
A necessary and sufficient condition for the implication (4.17) to be true is for there to exist a vector $\hat\lambda \in \mathbb{R}^{m+1}$ such that

$$\alpha = \begin{bmatrix} 1 & \beta \end{bmatrix} \hat\lambda = \lambda_0 + \beta\lambda \tag{4.23}$$

where

$$\hat\lambda = \begin{bmatrix} \lambda_0 \\ \lambda \end{bmatrix} \tag{4.24}$$

with $\lambda_0 \in \mathbb{R}$ and $\lambda \in \mathbb{R}^m$.

Many of you will hopefully recognize this as nothing more than the simple version of Ross' 1976 Arbitrage Pricing Theory (APT) [46]. Its derivation is given below.

Proof: Note that the condition in (4.18) indicates that if a portfolio x is in the null space of the matrix

$$\begin{bmatrix} \mathbf{1}^\mathbf{T} \\ \beta^T \end{bmatrix},$$

then it also must be orthogonal to α. Another way to state this is that

$$\alpha \in \left(\mathbf{N}\left(\begin{bmatrix} \mathbf{1}^\mathbf{T} \\ \beta^T \end{bmatrix} \right) \right)^\perp \tag{4.25}$$

where $\mathbf{N}(\cdot)$ is the null space. But, using the null and range space relationship, this means that

$$\alpha \in \mathbf{R}\left(\left[\begin{array}{c} \mathbf{1^T} \\ \beta^T \end{array}\right]^T\right).$$

(4.26)

Thus, there exists a vector $\hat{\lambda} \in \mathbb{R}^{m+1}$ such that $[\, \mathbf{1} \quad \beta \,]\hat{\lambda} = \alpha$. □

We will refer to this result, and equation (4.23) in particular, as the *Return APT*.

4.2.4 Interpretations

The vector $\hat{\lambda}$ is the key quantity in this absence of arbitrage condition. Let's try to get a bit of intuition about these mystical λ's.

4.2.4.1 Market Price of Risk

The Return APT equation for a one-factor model $r = \mu dt + \sigma dz$ is

$$\mu = [\, \mathbf{1} \quad \sigma \,]\hat{\lambda} = \lambda_0 + \sigma\lambda_1.$$

(4.27)

Let's analyze this equation. On the left side is the expected return or drift rate (μ) and on the right side is the volatility (σ). Hence, this equation relates volatility to expected return.

A better way to think of σ is not as volatility, but as the amount of the risk factor dz that affects the return. If σ is zero, then you are not exposed to the risk dz at all. If it is large, then you have purchased a lot of that risk. Therefore, λ_1 tells you how much your expected return is increased (assuming λ_1 is positive) for each unit of the risk dz to which you are exposed. In this sense, it is your "reward" per unit of the risk dz. For this reason, λ_1 is called the "market price of risk" because it is, in some sense, the price that you receive for holding a single unit of the risk dz. (It also goes by the names "factor price" and "factor premium," but personally I think "factor reward" would have been a better name!) This is a nice and very intuitive interpretation for λ_1.

4.2.4.2 Market Price of Time

We also have this pesky λ_0 to deal with. It is not tied to any risk factor dz. In fact, if we don't take on any risk (i.e., $\sigma = 0$), then $\mu = \lambda_0$. Intuitively, if we don't have any risk, then we should be earning the risk-free rate of return. Hence, we might guess that $\lambda_0 = r_0$, where r_0 denotes the risk-free rate of interest. This is in fact correct. But, let's justify this through a slightly different argument that will also lead us to a nice interpretation for λ_0.

Let us consider a risk-free asset (or risk-free bond), B. Its factor model is

$$\frac{dB}{B} = r_0 dt$$

(4.28)

where r_0 is a constant annualized risk-free rate of interest. Since it must satisfy our Return APT equation, we have

$$r_0 = \lambda_0 + 0(\lambda_1) = \lambda_0 \tag{4.29}$$

which tells us that $\lambda_0 = r_0$. If the λ's are the market prices of risky factors, then we may interpret λ_0 as the reward for the time factor dt or the "market price of time." Intuitively, this makes perfect sense, since the risk-free rate is the reward for taking on time and nothing else.

As a terminology note, I will often not take the effort to distinguish between λ_0 as the "market price of time" versus the other "market prices of risk." Rather, I will often refer to them all together, including λ_0, as "market prices of risk."

Now all the λ's should make sense. They relate the different factors dt, dz, etc. to how much we are rewarded for exposure to those factors. In a market with no arbitrage, every factor has a market price. Your return is just given by looking at how many units of each factor you are exposed to, and multiplying each by its respective market price and adding them up. Quite simple, right?

This is the basic interpretation of the concept of "market price of risk". Don't forget it. It is very helpful for your intuition.

4.2.5 A Problem with Returns

Using returns to model assets has a disadvantage. There are contracts that involve no up-front cost or price. Hence, they don't have a well defined return, since by definition a return involves dividing by the price, which is zero. A futures contract is an example of this, since the mark to market mechanism resets the value of a contract to zero every day. Therefore, in the above framework, a futures contract must be dealt with as a special case. I don't like special cases, so below we will reformulate absence of arbitrage conditions in terms of prices and portfolio value changes rather than returns. This eliminates the need for special cases.

4.3 FACTOR APPROACH USING PRICES AND VALUE CHANGES

I just mentioned that there can be some difficulties in dealing with returns. Therefore, in this section we will reformulate the factor approach to arbitrage pricing by working with prices and value changes rather than returns. In this case, it will be okay to have an asset with a zero price.

4.3.1 Prices, Value Changes, and Arbitrage

Let's return to our arbitrage portfolio and reformulate it in terms of prices and portfolio value changes. In our original argument, we let $r \in \mathbb{R}^n$ be the returns of assets and specified the dollar amount invested in each asset by a vector $x \in \mathbb{R}^n$.

This time we will specify a vector of prices $\mathcal{P} \in \mathbb{R}^n$, the profit/loss or change in portfolio value from holding a single unit of each tradable $d\mathcal{V} \in \mathbb{R}^n$, and shares or units of each asset held $y \in \mathbb{R}^n$.

By the change in portfolio value from holding a single unit of each tradable over the time period, I mean that if you acquire one unit of an asset at the beginning of the period, at the end of the period your portfolio value will have changed by an amount exactly equal to the corresponding element of $d\mathcal{V}$. Perhaps said more concretely, $d\mathcal{V}$ is the profit/loss vector over the period per unit held of the tradable assets. We assume that this change in value per unit follows the factor model,

$$d\mathcal{V} = \mathcal{A}dt + \mathcal{B}dz \tag{4.30}$$

where $\mathcal{A} \in \mathbb{R}^n$, $\mathcal{B} \in \mathbb{R}^{n\times m}$, and $dz \in \mathbb{R}^m$. The simple cash flow diagram is given in Figure 4.2.

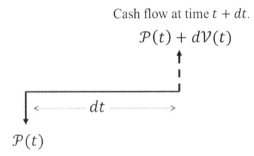

Cash flow at time $t + dt$.

$$\mathcal{P}(t) + d\mathcal{V}(t)$$

$$dt$$

$$\mathcal{P}(t)$$

Figure 4.2   Cash flow diagram with prices $\mathcal{P}$ and value changes $d\mathcal{V}$.

Note a couple of things about this cash flow modeling. We separate the prices in $\mathcal{P}(t)$, which represent the amounts that we have to pay right now for each single unit, from the changes in value (or profit/loss), $d\mathcal{V}$, that result from holding each unit. What is the purpose of separating these?

In the case of purchasing shares of a stock that doesn't pay a dividend, the price of a single unit is the price per share $S(t)$, while the profit/loss or change in value of what you hold is just the change in the price per share dS. That is, in this case, we simply have that $\mathcal{P} = S(t)$, and $d\mathcal{V} = dS$. There doesn't appear to be a need to distinguish between the change in price $d\mathcal{P}$ and the profit/loss from holding a single share $d\mathcal{V}$.

However, in the case of futures contracts, it doesn't cost anything to enter into a futures contract! Thus, the "price" of a unit (or contract) is 0, and we have $\mathcal{P} = 0$. By the mark to market mechanism of futures contracts, at the beginning of each day, the futures price f is adjusted so that the "price" of entering into a contract remains 0. However, despite the fact that the cost to enter into a contract is always reset to 0, the profit/loss from holding a single futures contract is given by the change in the futures price. That is, $d\mathcal{V} = df$. This is the essence of marking to market. In this case, the futures price is not the price of a contract, nor the price of anything that you can buy right now. But, it does determine the profit/loss, or the change in value of your portfolio. Thus, for situations such as this, it is advantageous to think separately about the price per unit of the asset, $\mathcal{P}$, and the change in the value of your portfolio from holding that unit, $d\mathcal{V}$.

Now that we understand the reasoning for this representation of a portfolio and tradable assets, let's construct our arbitrage portfolio and derive absence of arbitrage conditions.

4.3.2 Profit/Loss and Arbitrage

In this case, the profit/loss on the portfolio is given by the sum of the product of the number of units of each asset held, y, with the profit/loss from holding each unit $d\mathcal{V}$,

$$y^T(d\mathcal{V}) = y^T(\mathcal{A}dt + \mathcal{B}dz) = (y^T\mathcal{A})dt + (y^T\mathcal{B})dz, \tag{4.31}$$

Therefore, an arbitrage portfolio can be defined as follows.

(★) An Arbitrage Portfolio
An arbitrage portfolio is a specification of the number of units held of each asset $y \in \mathbb{R}^n$ *that satisfies the three equations*

$$y^T\mathcal{P} = 0 \quad \textit{No cost.}$$
$$y^T\mathcal{B} = 0 \quad \textit{No risk.}$$
$$y^T\mathcal{A} > 0 \quad \textit{Profit.}$$

Of course, we want to eliminate arbitrages, so we would like the following implication to hold, which once again is not in a form that is terribly useful.

(★) A (not very useful) Necessary Absence of Arbitrage Condition
If there is no arbitrage, then the following implication must be true,

$$\left. \begin{array}{ll} y^T\mathcal{P} = 0 & \textit{No cost} \\ y^T\mathcal{B} = 0 & \textit{No risk} \end{array} \right\} \Rightarrow y^T\mathcal{A} = 0 \quad \textit{No profit/loss,} \tag{4.32}$$

which can be written in matrix form as $\begin{bmatrix} \mathcal{P}^T \\ \mathcal{B}^T \end{bmatrix} y = 0 \Rightarrow \mathcal{A}^T y = 0.$

As in the case of returns, we can convert this to a dual condition that is useful.

(★★) Price APT: A Useful Necessary Absence of Arbitrage Condition
A necessary condition for no arbitrage is for there to exist a $\hat{\lambda} \in \mathbb{R}^{m+1}$ *such that*

$$\mathcal{A} = \begin{bmatrix} \mathcal{P} & \mathcal{B} \end{bmatrix} \hat{\lambda} = \mathcal{P}\lambda_0 + \mathcal{B}\lambda \tag{4.33}$$

where $\hat{\lambda} = \begin{bmatrix} \lambda_0 \\ \lambda \end{bmatrix}$ *with* $\lambda_0 \in \mathbb{R}$ *and* $\lambda \in \mathbb{R}^m$.

We refer to the above result, and equation (4.33) in particular, as the *Price APT*.

What is the relationship between the Return APT and the Price APT? Of course, they are basically equivalent. The λ's in both cases are the same, and have the same interpretation. They are market prices of risk. The main difference is that the price approach uses shares or units of the asset to describe the portfolio, not dollar amounts. This is really a superficial difference, but it is easier to understand some pricing situations, such as futures contracts, in terms of the Price APT rather than the Return APT.

4.4   TWO STANDARD EXAMPLES

To round out this chapter, we provide two examples to help us gain some intuition into the Return and Price APT equations. The first example considers a portfolio with a stock and a bond, where the stock provides information about the market price of risk for the risky factor.

4.4.1   A Stock under Geometric Brownian Motion

Assume that a stock follows a geometric Brownian motion and there is a bond that earns the risk-free rate, r_0. Their price dynamics are given as

$$dB = r_0 B dt, \tag{4.34}$$
$$dS = \mu S dt + \sigma S dz. \tag{4.35}$$

What do our absence of arbitrage conditions conclude from these tradable assets?

4.4.1.1   Using the Return APT

In terms of returns, we can write that

$$\frac{dB}{B} = r_0 dt, \tag{4.36}$$
$$\frac{dS}{S} = \mu dt + \sigma dz. \tag{4.37}$$

Hence, applying the absence of arbitrage Return APT equation (4.23) indicates there must exist a $\hat{\lambda} = [\lambda_0, \ \lambda_1]^T \in \mathbb{R}^2$ such that $\alpha = [\ \mathbf{1} \ \ \beta\ ]\hat{\lambda}$, where in this case

$$\alpha = \begin{bmatrix} r_0 \\ \mu \end{bmatrix}, \quad \mathbf{1} = \begin{bmatrix} 1 \\ 1 \end{bmatrix}, \quad \beta = \begin{bmatrix} 0 \\ \sigma \end{bmatrix}.$$

Thus, the Return APT condition is

$$\begin{bmatrix} r_0 \\ \mu \end{bmatrix} = \begin{bmatrix} 1 & 0 \\ 1 & \sigma \end{bmatrix} \begin{bmatrix} \lambda_0 \\ \lambda_1 \end{bmatrix}. \tag{4.38}$$

Solving for λ_0 and λ_1 gives

$$\lambda_0 = r_0 \tag{4.39}$$
$$\lambda_1 = \frac{\mu - r_0}{\sigma}. \tag{4.40}$$

These equations for the market price of time λ_0 and the market price of risk λ_1 should align with your intuition. The market price of time λ_0 is the reward for time, which should equal the risk-free rate, r_0. The market price of risk λ_1 is the reward per unit of the risky factor dz. In the stock, our overall reward (over the risk-free rate) is $\mu - r_0$, and we are holding σ units of the factor dz. Thus, our reward per unit of the risky factor is $\frac{\mu - r_0}{\sigma}$.

4.4.1.2   Using the Price APT

We can derive the same results using prices, value changes, and the Price APT. In this case, the price of each share of stock is $S(t)$, and the price per bond is $B(t)$. If we hold a unit of stock or bond, the profit/loss for each asset over each time period is just given by the change in their prices. That is,

$$dB = r_0 B dt, \tag{4.41}$$
$$dS = \mu S dt + \sigma S dz. \tag{4.42}$$

So, mapping this to the portfolio representation in the Price APT gives

$$\mathcal{P} = \begin{bmatrix} B \\ S \end{bmatrix}, \quad d\mathcal{V} = \begin{bmatrix} dB \\ dS \end{bmatrix}, \quad \mathcal{A} = \begin{bmatrix} r_0 B \\ \mu S \end{bmatrix}, \quad \mathcal{B} = \begin{bmatrix} 0 \\ \sigma S \end{bmatrix}.$$

Using these designations, the Price APT equation says that there must exist a $\hat{\lambda} = [\lambda_0, \ \lambda_1]^T \in \mathbb{R}^2$ such that $\mathcal{A} = [\ \mathcal{P} \ \mathcal{B} \ ]\hat{\lambda}$ or

$$\begin{bmatrix} r_0 B \\ \mu S \end{bmatrix} = \begin{bmatrix} B & 0 \\ S & \sigma S \end{bmatrix} \begin{bmatrix} \lambda_0 \\ \lambda_1 \end{bmatrix}. \tag{4.43}$$

Solving for λ_0 and λ_1 yields

$$\lambda_0 = r_0, \quad \lambda_1 = \frac{\mu - r_0}{\sigma},$$

as expected.

4.4.2   Futures Contracts

One can enter into a futures contract without paying any money. This means that a futures contract is a special case in our setup, and is difficult to understand in the context of returns because its return is not defined (it has zero price!). It is more naturally considered in terms of prices and value changes, and thus in the context of the Price APT. Figure 4.3 shows the cash flow diagram for a futures contract.

Recall that the critical difference between futures contracts and forward contracts is that futures contracts are marked to market and settled daily. This means that they always begin the day with a zero price. At the end of the day, the price change is settled, which means that the amount df, which is equal to the change in the futures prices that day, is either added to or deducted from your account

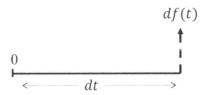

Figure 4.3 Cash flow diagram for futures contract.

for each contract you hold. Hence, the change in value of your portfolio for each contract you hold is directly equal to the change in the futures price, $d\mathcal{V} = df$.

Therefore, a futures contract can have a price of zero while the profit/loss from holding each contract is given by

$$d\mathcal{V} = df = \mu f\, dt + \sigma f\, dz. \tag{4.44}$$

Let's consider a market with a bond and a futures contract and see what our Price APT equations imply. The relevant quantities can be written as

$$\mathcal{P} = \begin{bmatrix} B \\ 0 \end{bmatrix}, \quad d\mathcal{V} = \begin{bmatrix} dB \\ df \end{bmatrix}, \quad \mathcal{A} = \begin{bmatrix} r_0 B \\ \mu f \end{bmatrix}, \quad \mathcal{B} = \begin{bmatrix} 0 \\ \sigma f \end{bmatrix}.$$

Now, let's apply the Price APT equation $\mathcal{A} = \begin{bmatrix} \mathcal{P} & \mathcal{B} \end{bmatrix} \hat{\lambda}$ with $\hat{\lambda} = [\lambda_0,\ \lambda_1]^T \in \mathbb{R}^2$, or

$$\begin{bmatrix} r_0 B \\ \mu f \end{bmatrix} = \begin{bmatrix} B & 0 \\ 0 & \sigma f \end{bmatrix} \begin{bmatrix} \lambda_0 \\ \lambda_1 \end{bmatrix}. \tag{4.45}$$

Solving this for λ_0 and λ_1 gives

$$\lambda_0 = r_0, \quad \lambda_1 = \frac{\mu}{\sigma}.$$

Once again, it is worthwhile to attempt to interpret the results. The bond represents pure exposure to the time factor dt, and thus the market price of time is just $\lambda_0 = r_0$ as we would have guessed.

On the other hand, the market price of risk λ_1 corresponding to the factor dz turns out to just be the expected rate of return of the futures price μ divided by the volatility σ. It is interesting to note that the risk-free rate does not appear in this market price of risk, as it did when calculating it from the stock in equation (4.40). Consider why this makes sense intuitively. A futures contract refers to the price that will be paid, in exchange for some asset, at a *fixed time* (the expiration) in the future. Hence, the actual payment cash flow corresponding to a futures contract doesn't move in time. It always stays at the expiration time of the contract, and thus has no exposure to the time factor dt. Because of this, we shouldn't expect it to be influenced by the market price of time, which is the risk-free rate. On the other hand, the profits and losses depend directly on the uncertainty in the futures price, which is driven by the risky factor dz. In this sense, the futures contract is a pure bet on the outcome of the risky factor. This explains why we can obtain the market price of risk for dz from the dynamics of a futures price without needing to know the risk-free rate.

4.5 SUMMARY

In this chapter, we first showed that stochastic differential equation models can be interpreted as linear factor models over an instantaneous time period dt. With this as the backdrop, we then set about deriving conditions in the context of factor models that must be satisfied for no arbitrage to exist. We provided two equivalent forms of the absence of arbitrage conditions, the Return APT and the Price APT equations. The key take-away was that to each risky factor driving a factor model for a tradable asset, there must exist a so-called market price of risk that measures the "reward" per unit of exposure to that risk. Moreover, we found that there must also exist a "market price of time" that provides a reward for exposure to time. Not surprisingly, the market price of time is just the risk-free rate of return.

The simple absence of arbitrage relationships established in the Return APT and Price APT equations will provide the basis for the rest of this book. They are that important! In fact, you will learn that just about every equation in derivative pricing is simply a statement of the APT equations derived in this chapter. In order to allow you to see this clearly, the next chapter is devoted to establishing a coherent and structured framework in which to apply these APT equations to the pricing of derivative securities. In the end, what you should take away is that the simple arbitrage ideas presented in this chapter are the underpinnings of derivative pricing theory.

EXERCISES

4.1 Consider a portfolio with dollar holdings $x = [\$5, -\$10, \$4]^T$ and returns following

$$
\begin{aligned}
r_1 &= 0.1dt + 0.3dz_1 \\
r_2 &= 0.2dt + 0.1dz_1 + 0.2dz_2 \\
r_3 &= 0.05dt.
\end{aligned}
$$

What is the factor model for the profit/loss on this portfolio?

4.2 Consider a portfolio with dollar holdings $x = [-\$5, \ \$7, -\$2]^T$ and returns following

$$
\begin{aligned}
r_1 &= 0.07dt \\
r_2 &= 0.15dt + 0.4dz_1 - 0.1dz_2 \\
r_3 &= 0.3dt - 0.6dz_1 + 0.3dz_2.
\end{aligned}
$$

What is the factor model for the profit/loss on this portfolio?

4.3 Consider a portfolio where $y = [2, -3, 6]^T$ is the number of units (shares) of each asset held. If the prices are $\mathcal{P} = [\$10, \$5.50, \$3]^T$, and the value change (profit/loss) per unit of the portfolio for each asset is

$$
\begin{aligned}
d\mathcal{V}_1 &= 0.05\mathcal{P}_1 dt \\
d\mathcal{V}_2 &= 0.1\mathcal{P}_2 dt + 0.4\mathcal{P}_2 dz_1 \\
d\mathcal{V}_3 &= 0.2\mathcal{P}_3 dt - 0.5\mathcal{P}_3 dz_1 - 0.1\mathcal{P}_3 dz_2,
\end{aligned}
$$

what is the factor model for the total value change of this portfolio over time dt, and what is the cost of the portfolio?

4.4 Consider a portfolio where $y = [-5, 2, 4]^T$ is the number of units (shares) of each asset held. If the prices are $\mathcal{P} = [\$18, \$3, \$12]^T$, and the value change (profit/loss) per unit of the portfolio for each asset is

$$
\begin{aligned}
d\mathcal{V}_1 &= 0.1\mathcal{P}_1 dt \\
d\mathcal{V}_2 &= 0.2\mathcal{P}_2 dt - 0.4\mathcal{P}_2 dz_1 + 0.2\mathcal{P}_2 dz_2 \\
d\mathcal{V}_3 &= 0.3\mathcal{P}_3 dt + 0.1\mathcal{P}_3 dz_1 - 0.3\mathcal{P}_3 dz_2,
\end{aligned}
$$

what is the factor model for the total value change of this portfolio over time dt, and what is the cost of the portfolio?

4.5 Consider a portfolio with three assets. Let the first two be stocks, with prices $S_1 = \$8$ and $S_2 = \$5$, and assume that the third asset is a futures contract. Let the futures price be $f = \$10$. Assume that the portfolio consists of 8 shares of stock S_1, 4 shares of stock S_2, and -5 futures contracts. Let the processes for the two stocks and the futures price be

$$
\begin{aligned}
dS_1 &= 0.1S_1 dt - 0.6S_1 dz_1 + 0.1S_1 dz_2 \\
dS_2 &= 0.2S_2 dt - 0.2S_2 dz_1 - 0.2S_2 dz_2 \\
df &= 0.3f dt + 0.3f dz_1 + 0.4f dz_2.
\end{aligned}
$$

What is the factor model for the total value change of this portfolio over time dt, and what is the cost of entering into this portfolio?

4.6 Consider a portfolio with three assets. Let the first two be stocks, with prices $S_1 = \$3$ and $S_2 = \$12$, and assume that the third asset is a futures contract. Let the futures price be $f = \$7$. Assume that the portfolio consists of 2 shares of stock S_1, -3 shares of stock S_2, and 10 futures contracts. Let the processes for the two stocks and the futures price be

$$
\begin{aligned}
dS_1 &= 0.05S_1 dt + 0.2S_1 dz_1 - 0.1S_1 dz_2 \\
dS_2 &= 0.12S_2 dt - 0.1S_2 dz_1 + 0.3S_2 dz_2 \\
df &= 0.2f dt + 0.0f dz_1 - 0.2f dz_2.
\end{aligned}
$$

What is the factor model for the total value change of this portfolio over time dt, and what is the cost of entering into this portfolio?

4.7 Consider a stock with price dynamics,

$$dS = 0.15Sdt + 0.3Sdz.$$

If the risk-free rate is $r_0 = 2\%$, determine the market price of time λ_0 and the market price of risk λ, corresponding to dz.

4.8 Consider a stock whose price follows

$$dS = 0.22Sdt + 0.4Sdz.$$

Also, assume that a futures contracts is traded on this stock. If the futures price follows

$$df = 0.2fdt + 0.4fdz,$$

then what is the absence of arbitrage risk-free rate in this market?

4.9 Consider a stock whose price follows

$$dS = 0.15Sdt + 0.3Sdz,$$

and assume that the risk-free rate is $r_0 = 3\%$. Consider a futures contract on the stock with corresponding futures price dynamics,

$$df = \mu_f fdt + 0.3fdz.$$

If there is no arbitrage, what must be the drift rate μ_f of the futures price corresponding to a futures contract on the stock?

4.10 Factor Model based on Binary Factor.

Consider a stock whose price can experience binary movement either up or down. That is, over each time step Δt, the stock price $S(k)$ can either move up to $S(k+1) = uS(k)$ or down to $S(k+1) = dS(k)$ with the probability structure

$$S(k+1) = \begin{cases} uS(k) & w.p. \ p \\ dS(k) & w.p. \ 1-p. \end{cases} \tag{P4.1}$$

Let Z be a standard binary random variable

$$Z = \begin{cases} 1 & w.p. \ p \\ -1 & w.p. \ 1-p. \end{cases} \tag{P4.2}$$

Show that $\Delta S(k) = S(k+1) - S(k)$ can be written in a factor model form as

$$\Delta S(k) = \mathcal{A}(k)\Delta t + \mathcal{B}(k)Z(k), \tag{P4.3}$$

and find $\mathcal{A}(k)$ and $\mathcal{B}(k)$.

CHAPTER 5

Constructing a Factor Model Pricing Framework

THIS chapter constructs the basic framework that we will use in the rest of the book for the modeling and pricing of derivative securities. We begin by classifying the quantities and variables involved in creating a mathematical model for derivative pricing. Specifically, we emphasize the key distinction between quantities that are tradable, and can be part of a portfolio, and those that are not. In particular, a derivative is just a specially designated tradable security that we would like to price.

To facilitate the derivation of derivative pricing equations, we focus on the creation of a special table that captures all the relevant information pertaining to the tradable assets. We call such a table a *tradable table*. With this information in hand, we proceed to apply and solve the Price APT equation from the previous chapter, the end result of which is the absence of arbitrage equation for the derivative security of interest.

Before finishing the chapter, the entire process of modeling, constructing a tradable table, and finally applying the Price APT is formalized into a three step procedure. In this way, the end goal of this chapter is to supply you with a detailed road map to guide you through the modeling and pricing of any derivative security.

5.1 CLASSIFICATION OF QUANTITIES

In this book, we are interested in the pricing of *derivative* securities. However, to understand that pricing theory, it is best to first understand the general modeling paradigm.

In our modeling framework, we will classify all quantities into three (possibly overlapping) categories. They are *factors*, *underlying variables*, and *tradables*.

5.1.1 Factors

Factors are the most basic sources of randomness in our models. In general, they are the driving Brownian motions $(z(t))$ and Poisson processes $(\pi(t))$. Furthermore, as you know, I like to think of time as a special factor. Thus, in our models, the factors will show up as the dz, $d\pi$, and dt terms.

5.1.2 Underlying Variables

Underlying variables are relevant quantities used to model and describe the assets that trade in a market. They are modeled as functions of the factors. For example, an interest rate could be an underlying variable. A stock price $S(t)$ could also be an underlying variable. In general, underlying variables play the role of the "known" quantities that we seek to express the prices of derivative securities in terms of. That is, there is an implicit assumption that underlying variables and the factor models that drive them are completely known and can be used to price assets.

5.1.3 Tradables

Tradables are the quantities that you can actually trade and include in a portfolio. They are modeled as functions of the underlying variables. Some of the time, the functional relationship between the tradable and an underlying variable is trivial. For example, a stock price $S(t)$ can be an underlying variable and also a tradable. In other cases, the tradable is a more complicated function of an underlying variable. For example, an interest rate $r(t)$ can be an underlying variable, but it is not tradable. Instead, a bond $B(r(t), t)$ is tradable, and represented as a function of the underlying variable interest rate.

The basic quantities and how they relate to each other are shown in Figure 5.1. First, factors are used to create models of the underlying variables. Then, the tradables are represented as functions of the underlying variables.

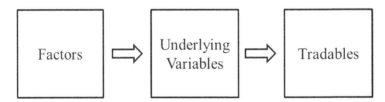

Figure 5.1 Modeling paradigm.

5.1.4 Importance of Tradables

It is *extremely important* to be able to separate quantities that are tradable from those that are not. Why? Because tradables — and *only* tradables — are the quantities that must satisfy the absence of arbitrage relationships from the previous chapter. It is those absence of arbitrage relationships that will lead to our pricing formulas.

To help hammer this home, let's consider some examples to solidify the notion of what is tradable.

Example 5.1 (Stock Paying a Dividend)
Consider a stock that pays a dividend. The stock has price $S(t)$, and let's assume that it pays a continuous dividend at a rate of q. What this means is that by reinvesting the dividend back into the stock, if you started by purchasing one share of stock at time 0, by time t you would have accumulated e^{qt} shares.

What is the value of the tradable quantity? You might be tempted to say that the price of the stock $S(t)$ represents the value of the tradable. However, note that you cannot just purchase the "price of the stock." Instead, you must purchase a *share* of stock, and with this share you not only get the price of the stock but also the dividend. The point is that you can't decouple the price from the dividend. They come together and the entire package is the tradable that must satisfy the APT relationship from the previous chapter. The stock price, on the other hand, is not tradable and does not need to satisfy the APT condition.

Example 5.2 (Bonds and Interest Rates)
In the context of bonds, we often deal with interest rates. However, interest rates are not tradable! You cannot purchase an interest rate. What you can purchase are bonds that depend on interest rates. Thus, bonds are the tradables in this context, and interest rates are underlying variables. So, as tradables, bonds will be bound by the APT equation from the previous chapter, while interest rates will not.

Example 5.3 (Futures Contracts)
As a final example, consider a futures contract. Associated with a futures contract is a futures price f(t) which corresponds to the price an investor has agreed to pay, in exchange for an asset, at the expiration time T of the futures contract. Once again, we need to clearly separate what the tradable quantities are. Note that the futures price does not represent the value of a tradable quantity. That is, when you enter into a futures contract, you do not hold a tradable whose value is equal to the futures price $f(t)$. Instead, the tradable is the futures contract whose value, by the mark to market mechanism, is reset to zero each day. Now, the cash flows generated from holding a futures contract definitely depend on the futures price $f(t)$, and thus the futures price is an underlying variable, but it is not a tradable. The tradable is the actual futures *contract* with its associated cash flows from marking to market.

5.2 DERIVATIVES

In this book we are interested in pricing derivative securities. So, it is important to begin by defining what we mean by a *derivative security*.

Derivative Security: *A derivative security is a tradable whose value depends upon (or is a function of) other underlying variables. In this case, we say that the derivative security is derivative to the underlying variables.*

Note that this definition of a derivative does not particularly distinguish it from any other tradable! We typically model all tradables as functions of some underlying variables. Even when a variable is both a tradable and an underlying variable, as in the case of the price of a non-dividend paying stock $S(t)$, we can think of it as trivially being a function of itself.

It is important to emphasize that from the perspective of absence of arbitrage, there is no real distinction between a derivative and any other tradable. You will see that the most important distinction is that a derivative is "what you want to price" and that other tradables are assumed to be "already priced" in the market.

The fact that a derivative is what we "want to price," also means that the exact formula for the price of the derivative as a function of the underlying variables is unknown. In fact, one can state the goal of derivative pricing as the following.

Goal of Derivative Pricing: *Determine the price of the derivative security explicitly as a function of underlying variables.*

This is a bit abstract, so let's consider some concrete examples next.

5.2.1 Examples of Derivatives

The first example below involving European call and put options, is considered the "standard example" when introducing derivative securities. In fact, I will use this example throughout the rest of the chapter to explain various concepts. So, make sure that you understand European calls and puts before proceeding.

Example 5.4 (European Call and Put Options)
A European call (put) option is a derivative security that gives the holder the option to purchase (sell) a specified stock, for a specified price called the *strike price*, at a specified date called the *expiration date*. For example, a European call option on stock XYZ with strike price $10 and expiration in 1 month would give the holder the right, but not the obligation, to purchase stock XYZ in exactly 1 month for $10. Similarly, the holder of a European put option with strike price $10 and expiration 1 month would have the right, but not the obligation, to sell the stock in exactly 1 month for a price of $10.

In this case, both call and put options are derivatives because their value depends on the stock price of XYZ. For example, consider the European call option and assume that stock XYZ does not pay a dividend. That is, the price of stock XYZ, $S(t)$, is an underlying variable (and a tradable) and the value of the European call option is derivative to $S(t)$. To see this, note that if the price of XYZ is more than $10 at expiration, then the holder of the option should "exercise" the option and buy for the strike price of $10. If the price of XYZ is less than $10 at expiration, then the option to buy for $10 is worthless. In this way, the stock price of XYZ is an underlying variable for the call option. In an exactly analogous manner, a European put option is also derivative to the stock price of XYZ.

In the preceding example, stock XYZ is an underlying variable and also a tradable. That is, you can actually buy and sell stock XYZ. However, it is not always the case that the underlying variable must also be tradable. Consider the following example.

Example 5.5 (Interest Rate Cap)
An interest rate cap puts a ceiling on the rate of interest that has to be paid for each payment of a floating rate loan. Thus, if the cap is at 5% and interest rates move to 6%, the cap limits the current payment to a rate of 5%. Therefore, a cap is a derivative whose value depends on the interest rate. In this case, the interest rate, which is the underlying variable, is not tradable.

In general, derivative securities are quite flexible and so are the possible underlying variables. In fact, in certain markets, there are derivatives that depend on underlying variables such as temperature or even wind speed which are not even financially related!

5.3   FACTOR MODELS FOR UNDERLYING VARIABLES AND TRADABLES

To use the Price APT equation from the previous chapter, we need linear factor models for the value changes of tradables. In the previous chapter we saw that we can interpret stochastic differential equations as instantaneous factor models. Hence, in continuous time models, we would like stochastic differential equation models for tradables and underlying variables that are driven by the factors.

In general, there are two ways that we obtain factor models written as stochastic differential equations. The first is that we directly model the underlying variables or tradables as factor models. The second (and most important) is Ito's lemma.

5.3.1   Direct Factor Models

In many cases, we begin the modeling process by writing a factor model, or equivalently a stochastic differential equation, for an underlying variable. Examples of this include the geometric Brownian motion model of stock price movement,

$$dS = \mu S dt + \sigma S dz, \tag{5.1}$$

or perhaps the mean reverting Ornstein–Uhlenbeck model of Section 3.4.1 for an interest rate,

$$dr = -b(r - a)dt + cdz. \tag{5.2}$$

Thus, in these situations, we begin the modeling process by directly writing stochastic differential equations (or factor models) for underlying variables and/or tradables.

5.3.2 Factor Models via Ito's Lemma

The second method to obtain factor models is via Ito's lemma. Since tradables, and derivatives in particular, are functions of underlying variables, if we have a stochastic differential equation model for an underlying variable then we can use Ito's lemma to obtain a stochastic differential equation for the tradable.

This is a particularly important idea for derivative securities, since we assume that they are functions of underlying variables, but the exact functional relationship is unknown. In fact, as mentioned previously, the goal of derivative pricing is to find that exact functional relationship. So, we often begin by representing the derivative as an unknown function of the appropriate underlying variables, and then use Ito's lemma to obtain a factor model for this unknown function. To illustrate this mechanism, consider the two examples given next.

Example 5.6 (Call Option on a Stock)
Consider the example of a European call option on a stock following geometric Brownian motion as in (5.1). We then make the assumption that the call option is a function of the underlying stock price $S(t)$ and time t, and write this unknown function as $c(S(t), t)$. Assuming $c(S(t), t)$ satisfies the conditions of Ito's lemma, its application gives

$$dc = (c_t + \mu S c_S + \frac{1}{2}\sigma^2 S^2 c_{SS})dt + \sigma S c_S dz, \qquad (5.3)$$

which provides a stochastic differential equation factor model for the call option c.

Example 5.7 (Bond Depending on an Interest Rate)
Assume that we have modeled an interest rate $r(t)$ as an underlying variable and provided the Ornstein–Uhlenbeck stochastic differential equation model for it as

$$dr = -b(r - a)dt + cdz.$$

Since a bond is a function of interest rates, we assume a bond is a function of $r(t)$ and time t, as $B(r(t), t)$. In this case, an application of Ito's lemma gives

$$dB = (B_t - B_r b(r - a) + \frac{1}{2}c^2 B_{rr})dt + cB_r dz,$$

which provides us with our stochastic differential equation factor model for a bond.

The typical route to stochastic differential equations for tradables is shown in Figure 5.2. Once we have obtained factor models for all the tradables in the market, we are ready to apply absence of arbitrage considerations. An extremely useful method of organizing the required information is given next.

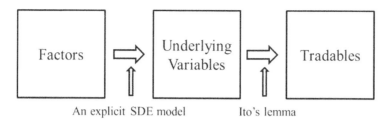

Figure 5.2 Typical route to obtaining factor models for tradables.

5.4 TRADABLE TABLES

We will approach the derivative pricing problem using the Price APT equation (4.33), rather than the Return APT (4.23). But note that everything in this book can also be done using the Return APT equation if desired.

To use the Price APT equation, we need two things: prices and factor models for value changes (profit/loss) of the tradables. The previous section indicated common approaches to obtaining the factor models for value changes. Combining these factor model stochastic differential equations with the prices of the tradables then encompasses all the required information for an application of the Price APT equation. When we gather all this essential information on the tradables into a single table, we call this a *tradable table*.

Tradable tables will play a key role throughout this book. They provide a structured and succinct way to display the information required to apply the Price APT, and allow us to systematically approach the various different derivatives we encounter.

5.4.1 Structure of a Tradable Table

A tradable table displays the prices and factor models for *all* the tradable securities in a market. That is, let $\mathcal{P}$ denote a vector of the prices of the tradable securities, and let

$$dV = \mathcal{A}dt + \mathcal{B}dz$$

be a vector/matrix representation of the factor model stochastic differential equations for the value changes of all the tradables. The tradable table is simply a table that aligns the prices contained in $\mathcal{P}$ with their associated value change factor models, as shown below.

$$
\begin{array}{c|c}
\text{Prices} & \text{Value Change Factor Models} \\
---- & ------------- \\
\mathcal{P} & dV = \mathcal{A}dt + \mathcal{B}dz
\end{array}
\tag{5.4}
$$

Tradable tables provide a uniform and universal method to display the key information required for the application of the Price APT equation. An example of constructing the tradable table for a market with a call option on a stock is given next.

Example 5.8 (Tradable Table for Bond, Stock, and Call Option)
Consider a market with a bond, a stock, and a call option on the stock. The factor model for a non-dividend paying stock is given by equation (5.1) and the factor model for the call option was derived via Ito's lemma in equation (5.3) of Example 5.3.2. Additionally, we consider a tradable risk-free bond earning the continuously compounded risk-free rate of interest r_0. The factor model for the risk-free bond is

$$dB = r_0 B dt$$

which corresponds to the formula for the value of the bond as $B(t) = B(0)e^{r_0 t}$ where $B(0)$ is the initial amount of money placed in the bond.

With these three securities — the bond, stock, and call option — as our tradables, we create a tradable table by first collecting their prices into the price vector

$$\mathcal{P} = \begin{bmatrix} B \\ S \\ c \end{bmatrix},$$

and then displaying this alongside the associated factor model stochastic differential equations for their value changes. The result is the following tradable table,

$$
\begin{array}{cc}
\text{Prices} & \text{Value Change Factor Models} \\
---- \; | & -------------
\end{array}
$$

$$
\begin{bmatrix} B \\ S \\ c \end{bmatrix} \quad
d \begin{bmatrix} B \\ S \\ c \end{bmatrix} = \begin{bmatrix} r_0 B \\ \mu S \\ (c_t + \mu S c_S + \frac{1}{2}\sigma^2 S^2 c_{SS}) \end{bmatrix} dt + \begin{bmatrix} 0 \\ \sigma S \\ \sigma S c_S \end{bmatrix} dz.
$$

Matching this with the generic tradable table representation in (5.4) leads to

$$
\mathcal{P} = \begin{bmatrix} B \\ S \\ c \end{bmatrix}, \quad
\mathcal{V} = \begin{bmatrix} B \\ S \\ c \end{bmatrix}, \quad
\mathcal{A} = \begin{bmatrix} r_0 B \\ \mu S \\ (c_t + \mu S c_S + \frac{1}{2}\sigma^2 S^2 c_{SS}) \end{bmatrix}, \quad
\mathcal{B} = \begin{bmatrix} 0 \\ \sigma S \\ \sigma S c_S \end{bmatrix}.
$$

Once a tradable table has been constructed, the next step is to apply the Price APT equation which enforces absence of arbitrage on the tradables.

5.5 APPLYING THE PRICE APT

When dealing with a tradable table for the purpose of pricing a derivative, the Price APT (4.33) is not applied equally across the entire table. Instead, we first separate what we call *marketed tradables* from the derivative that we would like to price. This separation distinguishes "what is already priced in the market" or the marketed tradables, from "what we want to price" which is the derivative. It also emphasizes the fact that derivative pricing is a form of *relative* pricing. That is, derivative pricing assumes that some securities are already priced in the market, and then proceeds to price the derivative security in a manner consistent with those already priced securities.

Separating out marketed tradables will also guide the order in which we apply and solve the Price APT equation (4.33). The Price APT applied to the marketed tradables is used to extract market prices of risk. Recall from the previous chapter that market prices of risk specify the "reward" per unit of exposure to each factor. Once determined, these market prices of risk are then plugged into the Price APT for the derivative, resulting in a pricing equation. This entire procedure is elaborated on next.

5.5.1   Relative Pricing and Marketed Tradables

Once we have set up the tradable table, we designate certain tradables as *marketed*. What we mean by a marketed tradable is a tradable that is "already priced" in the market.

You should contrast the marketed tradables with the derivative that we would like to price. The derivative is what we "want to price" using the information from the marketed tradables. As mentioned previously, this emphasizes the *relative pricing* nature of derivative pricing. We price a derivative security *relative* to the marketed tradables. That is, derivative pricing is a method of pricing a new tradable (the derivative) consistently with the existing prices of *marketed tradables.*

To see how this works, we separate the marketed tradables from the derivative, and place them all in a tradable table:

$$
\begin{array}{cc}
\text{Prices} & \text{Value Change Factor Models} \\
- - - - \quad | & - - - - - - - - - - - - - - - \\
\begin{bmatrix} \mathcal{P}_m \\ -- \\ \mathcal{P}_d \end{bmatrix} & d \begin{bmatrix} \mathcal{V}_m \\ -- \\ \mathcal{V}_d \end{bmatrix} = \begin{bmatrix} \mathcal{A}_m \\ -- \\ \mathcal{A}_d \end{bmatrix} dt + \begin{bmatrix} \mathcal{B}_m \\ -- \\ \mathcal{B}_d \end{bmatrix} dz.
\end{array}
$$

The tradables with the subscript m are the marketed tradables, and the tradable with the subscript d is the derivative we are pricing. The convention that we will use in this book is to always place the marketed tradables first in the tradable table, and then place the derivative last, separated from the marketed tradables with horizontal dashes.

5.5.2 Pricing the Derivative

Now we can use the Price APT to price the derivative. The Price APT equation says that to eliminate arbitrage, there must exist λ_0 and λ such that

$$\mathcal{A}_m = \mathcal{P}_m \lambda_0 + \mathcal{B}_m \lambda. \tag{5.5}$$
$$\mathcal{A}_d = \mathcal{P}_d \lambda_0 + \mathcal{B}_d \lambda. \tag{5.6}$$

To obtain the market prices of risk λ_0 and λ, we first *only* solve the Price APT for the *marketed tradables*. That is, we do not treat the equations (5.5) and (5.6) equally. Instead, initially we solve (5.5) to find the market prices of risk (often in terms of information from the marketed tradables). The resulting market prices of risk are then plugged into (5.6) which gives us the pricing equation for the derivative.

One may think of this procedure as first *calibration* to market data, followed by pricing the new derivative. Using marketed tradables to determine the market prices of risk is like calibrating parameters (market prices of risk) in our absence of arbitrage pricing model to known data (marketed tradables). Once our model is calibrated (the market prices of risk are determined), we can apply our absence of arbitrage model to price other tradables (derivatives) that also must satisfy the same absence of arbitrage relationship.

This discussion has been a bit abstract, so let's see it applied to an example.

Example 5.9 (Pricing Equation for a European Call Option)
Let's take the tradable table in Example (5.4.1) and use it to illustrate pricing. Assume that c is a European call option on the stock S. First, we designate the bond B and stock S as marketed tradables, and c as the derivative that we would like to price. That is, we separate the tradable table into the marketed tradables and the derivative:

$$
\begin{array}{cc}
\text{Prices} & \text{Value Change Factor Models} \\
- - - - \mid & - - - - - - - - - - - - - - - \\
\begin{bmatrix} B \\ S \\ -- \\ c \end{bmatrix} & d\begin{bmatrix} B \\ S \\ -- \\ c \end{bmatrix} = \begin{bmatrix} r_0 B \\ \mu S \\ -- \\ (c_t + \mu S c_S + \frac{1}{2}\sigma^2 S^2 c_{SS}) \end{bmatrix} dt + \begin{bmatrix} 0 \\ \sigma S \\ -- \\ \sigma S c_S \end{bmatrix} dz.
\end{array}
$$

Mapping this to the notation for a generic tradable table gives

$$\mathcal{P}_m = \mathcal{V}_m = \begin{bmatrix} B \\ S \end{bmatrix}, \quad \mathcal{A}_m = \begin{bmatrix} r_0 B \\ \mu S \end{bmatrix}, \quad \mathcal{B}_m = \begin{bmatrix} 0 \\ \sigma S \end{bmatrix}$$

for the marketed tradables, and

$$\mathcal{P}_d = \mathcal{V}_d = \begin{bmatrix} c \end{bmatrix}, \quad \mathcal{A}_d = \begin{bmatrix} (c_t + \mu S c_S + \frac{1}{2}\sigma^2 S^2 c_{SS}) \end{bmatrix}, \quad \mathcal{B}_d = \begin{bmatrix} \sigma S c_S \end{bmatrix}$$

for the derivative. The Price APT applied to the entire tradable table leads to

$$
\begin{bmatrix} r_0 B \\ \mu S \\ -- \\ c_t + \mu S c_S + \frac{1}{2}\sigma^2 S^2 c_{SS} \end{bmatrix} = \begin{bmatrix} B \\ S \\ -- \\ c \end{bmatrix} \lambda_0 + \begin{bmatrix} 0 \\ \sigma S \\ -- \\ \sigma S c_S \end{bmatrix} \lambda_1 \tag{5.7}
$$

where the top two rows correspond to the marketed tradables. When solving the Price APT, we first only use the top two equations for the marketed tradables,

$$
\begin{bmatrix} r_0 B \\ \mu S \end{bmatrix} = \begin{bmatrix} B \\ S \end{bmatrix} \lambda_0 + \begin{bmatrix} 0 \\ \sigma S \end{bmatrix} \lambda_1.
$$

Solving these equations for the market prices of risk results in

$$
\lambda_0 = r_0, \qquad \lambda_1 = \frac{\mu - r_0}{\sigma}.
$$

Finally, we use these calibrated market prices of risk in the third equation of (5.7) for our derivative to obtain

$$
c_t + \mu S c_S + \frac{1}{2}\sigma^2 S^2 c_{SS} = r_0 c + \frac{\mu - r_0}{\sigma}\sigma S c_S. \tag{5.8}
$$

Rearranging this leads to

$$
c_t + r_0 S c_S + \frac{1}{2}\sigma^2 S^2 c_{SS} = r_0 c, \tag{5.9}
$$

which is the absence of arbitrage Black–Scholes partial differential equation that the price of a call option must satisfy.

A diagram of the application of the Price APT to a tradable table containing marketed tradables and a derivative security is given in Figure 5.3.

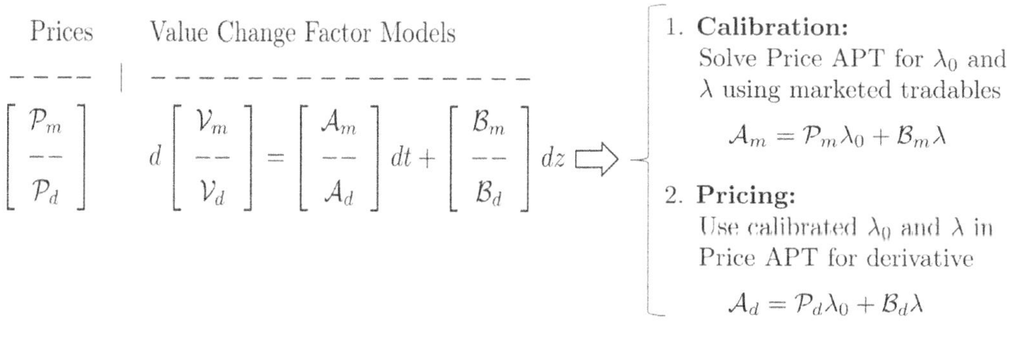

Figure 5.3 Application of the Price APT to a tradable table.

5.6 UNDERDETERMINED AND OVERDETERMINED SYSTEMS

In Example 5.5.2 above, everything was perfect! I had two equations for the marketed tradables (one for the bond and one for the stock) and two unknowns (the market price of time λ_0 and the market price of risk λ_1). Thus, I could uniquely solve the equations for λ_0 and λ_1.

Now, what if things aren't so perfect? That is, what if the system of equations arising from the Price APT for the marketed tradables is either underdetermined or overdetermined?

5.6.1 Underdetermined and Incompleteness

Assume that our marketed tradables are given by

$$dB = r_0 B dt \tag{5.10}$$
$$dS = \mu S dt + \sigma_1 S dz_1 + \sigma_2 S dz_2, \tag{5.11}$$

with corresponding tradable table:

$$
\begin{array}{cc}
\text{Prices} & \text{Value Change Factor Models} \\
- - - - \quad | & - - - - - - - - - - - - - - - \\
\begin{bmatrix} B \\ S \end{bmatrix} &
d\begin{bmatrix} B \\ S \end{bmatrix} = \begin{bmatrix} r_0 B \\ \mu S \end{bmatrix} dt + \begin{bmatrix} 0 & 0 \\ \sigma_1 S & \sigma_2 S \end{bmatrix} \begin{bmatrix} dz_1 \\ dz_2 \end{bmatrix}.
\end{array}
$$

In this case, the factors are dt, dz_1, and dz_2, and the Price APT equation is

$$
\begin{bmatrix} r_0 B \\ \mu S \end{bmatrix} = \begin{bmatrix} B & 0 & 0 \\ S & \sigma_1 S & \sigma_2 S \end{bmatrix} \begin{bmatrix} \lambda_0 \\ \lambda_1 \\ \lambda_2 \end{bmatrix}. \tag{5.12}
$$

We have three unknowns (λ_0, λ_1, and λ_2), but only two equations! This system of equations is underdetermined. Thus, we can't uniquely solve for the market prices of risk! (In this case, we can solve for $\lambda_0 = r_0$, but not uniquely for λ_1 or λ_2.)

What can we say in this situation and how should we think of this? Well, first we can say that if *any* solution exists (it doesn't have to be unique, we just need at least one solution to exist), then there is no arbitrage. This is guaranteed by the Price APT equations.

Let's assume that many solutions exist. For example, in the above equations there are multiple possible values for λ_1 and λ_2. Thus, there are many possible market prices of risk that satisfy the no-arbitrage condition. This just means that from the tradable assets in the market (B and S), we *cannot* uniquely infer the market prices of risk for dz_1 and dz_2. There are many possibilities, and all are arbitrage free.

The practical consequence of this is that if we are asked to price a new security that depends on dz_1 and/or dz_2, we will not be able to assign it a *unique* absence of arbitrage price. This is because the APT equation (in either Return or Price form) acts as a pricing equation. (This use will become clear in the following chapters.) However, it only provides a unique price if we have unique values for the market prices of risk which, in turn, provide a unique pricing equation. This situation is called an *incomplete* market.

Incomplete markets are common in practice, and you will see in subsequent chapters that to price derivative securities in incomplete markets, we must *select* values for market prices of risk that are not uniquely defined. Since market prices of risk relate risk to reward for various factors, selecting a value for a market price of risk is essentially the same as specifying how investors in the market trade off risk and return. Thus, in incomplete markets, some specification of the risk preferences of investors is needed to assign a unique price to derivative securities. Furthermore, this specification of risk preferences is captured by the selection of the market price of risk.

In subsequent chapters we will encounter this situation. In fact, you might want to refer back to this discussion when faced with pricing of derivatives in incomplete markets (see for example, jump diffusion models or stochastic volatility).

5.6.2 Overdetermined and Calibration

Now let's consider the opposite situation. Assume the Price APT set of equations for the marketed tradables is overdetermined. For example, let the marketed tradables be

$$dB = r_0 B dt \tag{5.13}$$
$$dS_1 = \mu_1 S_1 dt + \sigma_1 S_1 dz \tag{5.14}$$
$$dS_2 = \mu_2 S_2 dt + \sigma_2 S_2 dz, \tag{5.15}$$

with corresponding tradable table:

Prices   Value Change Factor Models

$$
d\begin{bmatrix} B \\ S_1 \\ S_2 \end{bmatrix} = \begin{bmatrix} r_0 B \\ \mu_1 S_1 \\ \mu_2 S_2 \end{bmatrix} dt + \begin{bmatrix} 0 \\ \sigma_1 S_1 \\ \sigma_2 S_2 \end{bmatrix} dz.
$$

In this case, the Price APT gives

$$
\begin{bmatrix} r_0 B \\ \mu_1 S_1 \\ \mu_2 S_2 \end{bmatrix} = \begin{bmatrix} B & 0 \\ S_1 & \sigma_1 S \\ S_2 & \sigma_2 S \end{bmatrix} \begin{bmatrix} \lambda_0 \\ \lambda_1 \end{bmatrix}, \tag{5.16}
$$

and there are three equations and only two unknowns (λ_0 and λ_1)! This system looks to be overdetermined!

Now, we know from the Price APT that for no arbitrage to exist there must be a solution to this set of equations. However, in general, for an overdetermined system of equations no solution will exist! What does this mean?

Well, the first thing it means is that strictly speaking, there is an arbitrage opportunity. But, the way this situation often plays out in practice is usually slightly different. In practice this situation often leads to some sort of best-fit *calibration* procedure.

Instead of declaring that an arbitrage exists, a trader will often just assume that the models being used for B, S_1 and S_2 are not perfect, and assign that as the reason that no solution exists. Thus, the trader will search for the λ_1 that best fits the absence of arbitrage equations (5.16) in some appropriate sense. Watch for this in situations such as term structure modeling where a single factor dz is used, but many tradables in the form of bonds of different maturities, exist. Again, you might want to return to this discussion after reading subsequent chapters.

5.7   THREE STEP PROCEDURE

To begin to wrap things up in this chapter, let's summarize the approach we have developed so far. In fact, in order to provide the most clarity into the process of obtaining derivative pricing equations, we will map the approach to the following three step procedure.

1. **Model and Classify Variables:** Model and identify the tradable assets, underlying variables, and factors. (See Figure 5.1.)

2. **Construct a Tradable Table:** Based on the prices and factor models for each tradable asset, construct a tradable table as in (5.4).

3. **Apply the Price APT:** The marketed tradables are used in the Price APT equation to solve for the market prices of risk (calibration), which are then substituted into the Price APT equation for the derivative (pricing). (See Figure 5.3.)

In the following chapters I will refer to these three steps often. Thus, make sure you are familiar with them. This three step procedure will be our road map to deriving the absence of arbitrage equations that various derivative securities must satisfy.

After obtaining an absence of arbitrage equation via the three step procedure, there is one final step left. The last step is to specify the unique payoff characteristics of the derivative under consideration and then to actually solve the pricing equation while respecting the payoff constraint.

5.8   SOLVING THE PRICE APT EQUATION FOR A SPECIFIC DERIVATIVE

The Price APT equation for a derivative security just enforces absence of arbitrage in the factor model stochastic differential equation that describes the derivative. However, it does not specify which derivative is being considered. In fact, two different derivative securities that depend on the same underlying variables in the same manner will result in the same absence of arbitrage equation!

If we think about it, this should make sense. All derivative securities should be absence of arbitrage, and hence should satisfy the Price APT absence of arbitrage equation. That is, the unique aspect of a derivative security is determined by its payoff characteristics, not the fact that it doesn't allow arbitrage. The payoff characteristics of a derivative usually play the role of a boundary condition for the Price APT equation.

Example 5.10 (Specifying European Call and Put Payoffs)
Recall the definitions of European call and put options, as explained in Example 5.2.1. Both calls and puts on the same stock $S(t)$ can be represented as some unknown function, $c(S(t), t)$ for the call and $p(S(t), t)$ for the put. Moreover, both calls and puts will lead to the same tradable table given by (5.7), and result in the same Price APT absence of arbitrage equation,

$$c_t + r_0 S c_S + \frac{1}{2}\sigma^2 S^2 c_{SS} = r_0 c, \tag{5.17}$$

where one needs to only replace c with p to obtain the equation for the put. What distinguishes a call from a put is the difference in their payoffs.

In the case of the call option, at the expiration time T, we know that if the stock price is greater than the strike price K, then the option saves us an amount $S(T) - K$ when exercised. Said another way, we could buy the stock for an amount K by using the call option, and then immediately sell in the market for $S(T)$, realizing a profit of $S(T) - K$. This implies that $c(S(T), T) = S(T) - K$ for $S(T) \geq K$. On the other hand, if $S(T)$ is below the strike price K at expiration, the option is worthless, or $c(S(T), T) = 0$ for $S(T) < K$. These two scenarios combine to yield the payoff of the call and can be succinctly written together as

$$c(S(T), T) = \max\{S(T) - K), 0\}.$$

In a similar manner, a put option with strike price K and expiration T will be worthless if $S(T) > K$, and benefit the holder an amount $K - S(T)$ if exercised when $S(T) \leq K$. Thus, the payoff of a put option $p(S(T), T)$ can be written as

$$p(S(T), T) = \max\{K - S(T), 0\}.$$

These distinct payoff characteristics for the call and put, when considered in conjunction with the Price APT equation, (5.17), will lead to a unique price for each.

Example 5.11 (Option "Moneyness")
There is common terminology that accompanies the payoff functions of call and put options, and is often referred to as the *moneyness* of the option. Given the current price of the stock S on which an option is written, if exercising the option right now would lead to a positive payoff (even if that is not allowed, as in a European option prior to expiration), the option is said to be "in-the-money." In a similar

manner, if immediate exercise would lead to a zero payoff, the option is said to be "out-of-the-money." If the current price S is equal to the strike price, then an option is "at-the-money."

For example, if the strike price of a call option is $K = \$10$, then any current price of $S > \$10$ would make the call option in-the-money, since the payoff would be equal to $\max\{S - K,\ 0\} = S - K > 0$. On the other hand, a current price of $S < \$10$ is considered out-of-the-money for a call option since it leads to the payoff of $\max\{S - K,\ 0\} = 0$. If $S = K = \$10$, then the option is said to be at-the-money. Similarly, a put option with the same strike is out-of-the-money for $S > \$10$, at-the-money for $S = \$10$, and in-the-money when $S < \$10$.

Payoff characteristics of derivatives can come in many forms. Sometimes they are very explicit, as in the example above for European calls and puts. In other cases, the holder of the derivative has a lot of flexibility in the manner and timing of receiving a payoff, and this can add considerable complexity. In any case, the payoff can very much affect the difficulty with which a solution to the absence of arbitrage equation can be found. In some cases, closed form solutions exist. We will explore a number of such cases in the chapters to follow. In many other situations, one must resort to numerical methods, which will be the subject of Chapter 9.

5.9 SUMMARY

In this chapter, we built a three step procedure for deriving the absence of arbitrage pricing equation for a derivative security based on the factor model approach. This systematic approach will serve as our road map as we traverse many derivative models in the chapters to come. To convince you of the power and generality of this approach, in the next chapter we tackle examples in equity derivatives, and in the following chapter we address interest rate and credit derivatives.

The real power of the factor model framework is that it provides a unified structure within which to address derivative pricing. By thinking broadly in terms of factors, underlying variables, and tradables, the superficial differences between equity, interest-rate, credit, and other types of derivatives are swept away, and the simple shared factor model foundation is revealed.

EXERCISES

5.1 Tradable or Not?

Decide whether the following are tradable or not.

(a) A non-dividend paying stock.

(b) The price of a stock that pays a continuous dividend.

(c) The futures price of a barrel of crude oil.

(d) The yield on a 10-year US Treasury note.

(e) A forward contract to deliver a bushel of wheat in 3 months for $40.

(f) A zero-coupon bond that pays $1 in 30 years.

5.2 Calculate the payoff value of a European call option at expiration T with strike price $K = \$20$ when $S(T)$ takes the values given below. In each case, indicate whether the option should be exercised or not, and whether it is in-the-money, at-the-money, or out-of-the-money. (See Example 5.8.)

(a) $S(T) = \$23$.

(b) $S(T) = \$20$.

(c) $S(T) = \$15$.

5.3 Calculate the payoff value of a European put option at expiration T with strike price $K = \$20$ when $S(T)$ takes the values given below. In each case, indicate whether the option should be exercised or not, and whether it is in-the-money, at-the-money, or out-of-the-money. (See Example 5.8.) .

(a) $S(T) = \$23$.

(b) $S(T) = \$20$.

(c) $S(T) = \$15$.

5.4 Consider an interest rate *cap* as described in Example 5.2.1. Write the payoff function of the cap applied to just a single interest payment (this is called a *caplet*). Assume the principal of the loan is $\$1,000,000$, and interest payments are made every half year. Assume that the interest rate for the next payment is denoted R, and is quoted using a yearly convention, so that the next half year interest payment will be computed by dividing the yearly interest rate R in half, and multiplying by the loan principal. If the strike rate of the cap is 5% as in Example 5.2.1, then what is the payoff function for this cap applied to the next interest payment?

5.5 A *floorlet* gives the holder a minimum rate of interest corresponding to a single interest payment. Consider the same setup as in Exercise 5.4, with a principal of $P = \$1,000,000$ and half year interest payments. If the strike rate is 5%, what is the payoff function of a floorlet?

5.6 Create a tradable table for a risk-free asset $B(t)$ that earns the risk-free rate r_0 and a futures contract with corresponding futures price $f(t)$ that follows

$$df = \mu f \, dt + \sigma f \, dz.$$

5.7 A Stock Paying Continuous Dividends.

Consider a stock whose price $S(t)$ follows

$$dS = \mu S dt + \sigma S dz. \tag{P5.1}$$

Assume that the stock pays a continuous dividend at a rate of q. Also assume that a risk-free bond exists and satisfies

$$dB = r_0 B dt. \tag{P5.2}$$

If $c(S, t)$ is a European call option on the stock, then identify the factors and underlying variables, and create a tradable table. (Hint: Consider Example 5.1.4 and let $v(t)$ be the value of buying and holding a single share. Argue that v satisfies $dv = (\mu + q)v dt + \sigma v dz$.)

5.8 A Single Factor Short Rate Model.

Let $r_0(t)$ denote the so-called short rate of interest, and let

$$dr_0 = a dt + b dz. \tag{P5.3}$$

Assume that zero-coupon bonds of maturity T are tradables, with price at time t denoted by $B(r_0, t | T)$. That is, let these bonds be a function of the short rate r_0 and time t. Furthermore, assume that a money market account exists that satisfies

$$dB_0 = r_0(t) dB_0 dt. \tag{P5.4}$$

In this model, identify the factors, underlying variables, and tradables, and create a tradable table.

5.9 Asian Options.

An Asian option is a type of option that involves the average price of a stock over some period of time, given by $A(t) = \frac{1}{t} \int_0^t S(\tau) d\tau$. Consider an average strike Asian call option. This option allows the holder to purchase a share of the stock at expiration for the average price of the stock from the initiation of the option at time 0 until expiration at time T. To model this situation, consider using the integral $I(t) = \int_0^t S(\tau) d\tau$ as an underlying variable. Write a factor model for $I(t)$, and write the payoff of the average strike Asian call as a function of $I(t)$.

5.10 Let $S(t)$ be a non-dividend paying stock following $dS = \mu S dt + \sigma S dz_1$, and let $r_0(t)$ represent the short rate of interest earned by investing in a money market account that follows $dB_0 = r_0 B_0 dt$. Moreover, assume that the short rate is random and satisfies $dr = -b(r - a)dt + \eta dz_2$ where the correlation coefficient between dz_1 and dz_2 is ρ. Let $c(r_0(t), S(t), t)$ be a derivative security that depends on both $r_0(t)$ and $S(t)$. Use Ito's lemma to find the factor model that c follows. Create a tradable table.

5.11 Put-Call Parity.

Consider a non-dividend paying stock and associated call and put options with expiration time T. Show that the payoff function of a long position in one call option plus a short position in one put option plus a long position in K zero-coupon bonds (where each bond pays \$1 at the expiration time T) has exactly the same payoff as one share of the stock. Use this fact to argue that the current price at time t of the stock should equal the price of the call option minus the price of the put plus the price of K zero-coupon bonds. That is, argue that

$$S(t) = c(S(t), t) - p(S(t), t) + KB(t|T)$$

where $S(t)$ is the price of the stock, $c(S(t), t)$ is the price of the call option, $p(S(t), t)$ is the price of the put, and $B(t|T)$ is the current price of a zero-coupon bond that pays \$1 at T.

CHAPTER 6

Equity Derivatives

$\mathbf{T}$ HE factor model approach to absence of arbitrage pricing, as detailed in the previous chapter, is one of the most transparent and direct routes to deriving the equations that govern derivatives. In this chapter, we will explore its application to equity derivatives.

Our take-off point will be the three step procedure outlined in Section 5.7 of the previous chapter. Each equity derivative model will be approached systematically by walking carefully through each step. The purpose of this is to encourage you to think about derivative pricing in terms of factors, underlying variables, and tradables, which should allow you to abstract away from the details of any particular model and focus instead on the core absence of arbitrage fundamentals.

Along the way, we will take the time to introduce closed form solutions to various derivatives that arise from these models. This will allow us to become familiar with some of the standard pricing formulas for derivatives and their properties. We will begin where every respectable presentation of equity derivatives should begin — with the celebrated Black–Scholes model.

6.1   BLACK–SCHOLES MODEL

Black and Scholes [5] started everything with this model and the accompanying closed form solution for European call and put options. Elements of this model were considered in the examples throughout the previous chapter, but here we will carefully step through the entire process using the three step procedure.

Step 1: Model and Classify Variables
The standard Black–Scholes setup involves a bond $B(t)$ earning a continuously compounded annualized risk-free rate, r_0. This corresponds to the bond following the model

$$dB = r_0 B dt. \tag{6.1}$$

The market also contains a non-dividend paying stock with price $S(t)$ that follows a geometric Brownian motion (GBM),

$$dS = \mu S dt + \sigma S dz, \tag{6.2}$$

where μ is the expected rate of return and σ is the volatility.

We then consider a derivative on the stock. Generically, we call this derivative c and assume that its price process is a function of the price of the stock and time $c(S(t),t)$. (If it helps to keep something concrete in mind, then you are welcome to assume that c stands for the price of a European call option.) Our goal is to derive the absence of arbitrage equation that $c(S(t),t)$ must satisfy. To do this, we need a linear factor model for $c(S,t)$. If we assume that it is twice continuously differentiable, by Ito's lemma applied to $c(S,t)$ we obtain

$$dc = (c_t + \mu S c_S + \frac{1}{2}\sigma^2 S^2 c_{SS})dt + \sigma S c_S dz. \qquad (6.3)$$

An important aspect of Step 1 is to classify the relevant quantities in our model into factors, underlying variables, and tradables. In this case, we classify the variables as

Tradables	Underlying Variables	Factors
B, S, c	r_0, S	dt, dz

Note that here S is both an underlying variable and a tradable. This won't always be the case; however this often happens in equity derivatives. Now we can move to Step 2 and construct a tradable table.

Step 2: Construct a Tradable Table

Our task in constructing the tradable table is to gather and organize all the key information regarding the prices and factor models for the tradables. Moreover, we also need to distinguish the marketed tradables (tradables that are "already priced") from the derivative ("what we want to price").

As a reminder, the tradable table will always be organized as:

$$
\begin{array}{cc}
\text{Prices} & \text{Value Change Factor Models} \\
\hline
\begin{bmatrix} \mathcal{P}_m \\ -- \\ \mathcal{P}_d \end{bmatrix} &
d\begin{bmatrix} \mathcal{V}_m \\ -- \\ \mathcal{V}_d \end{bmatrix} =
\begin{bmatrix} \mathcal{A}_m \\ -- \\ \mathcal{A}_d \end{bmatrix} dt +
\begin{bmatrix} \mathcal{B}_m \\ -- \\ \mathcal{B}_d \end{bmatrix} dz
\end{array}
$$

where $\mathcal{P}_m$ and $d\mathcal{V}_m = \mathcal{A}_m dt + \mathcal{B}_m dz$ contain the the prices and factor models for the *marketed* tradables, and $\mathcal{P}_d$ and $d\mathcal{V}_d = \mathcal{A}_d dt + \mathcal{B}_d dz$ capture the price and factor model for the *derivative*. To distinguish between these in tradable tables, information for the marketed tradables will lie above the horizontal dashes, and the information for the derivative will lie below.

Composing this information for the tradables B, S, and c, results in the following completed tradable table:

$$
\begin{array}{cc}
\text{Prices} & \text{Value Change Factor Models}
\end{array}
$$

$$
\begin{bmatrix} B \\ S \\ -- \\ c \end{bmatrix} \quad d \begin{bmatrix} B \\ S \\ -- \\ c \end{bmatrix} = \begin{bmatrix} r_0 B \\ \mu S \\ ----- \\ c_t + \mu S c_S + \frac{1}{2}\sigma^2 S^2 c_{SS} \end{bmatrix} dt + \begin{bmatrix} 0 \\ \sigma S \\ -- \\ \sigma S c_S \end{bmatrix} dz.
$$

In this case, the tradables above the line (B and S) are marketed, while the tradable below the line (c) is the derivative to be priced. The tradable table contains all the information we need to apply the Price APT equation, which we proceed to do in Step 3.

Step 3: Apply the Price APT

In this step, we apply the Price APT equation using the information in the tradable table above. That is, we set up the absence of arbitrage equations

$$
\begin{aligned}
\mathcal{A}_m &= \mathcal{P}_m \lambda_0 + \mathcal{B}_m \lambda \\
\mathcal{A}_d &= \mathcal{P}_d \lambda_0 + \mathcal{B}_d \lambda
\end{aligned}
$$

where the top equation is for the marketed tradables and the bottom equation is for the derivative. In this case, we obtain:

$$
\begin{bmatrix} r_0 B \\ \mu S \\ --- \\ c_t + \mu S c_S + \frac{1}{2}\sigma^2 S^2 c_{SS} \end{bmatrix} = \begin{bmatrix} B \\ S \\ -- \\ c \end{bmatrix} \lambda_0 + \begin{bmatrix} 0 \\ \sigma S \\ -- \\ \sigma S c_S \end{bmatrix} \lambda_1. \tag{6.4}
$$

Recall that we proceed through these Price APT equations in two steps. The first step is to use the *marketed* tradables to solve for the market prices of risk. This can be considered the calibration phase. That is, we determine λ_0 and λ_1 using the first two equations which lie above the horizontal dashes. This gives

$$
\lambda_0 = r_0 \qquad \lambda_1 = \frac{\mu - r_0}{\sigma}. \tag{6.5}
$$

The second step is to substitute these values for λ_0 and λ_1 into the last equation in (6.4), which corresponds to the derivative c. This yields

$$
c_t + \mu S c_S + \frac{1}{2}\sigma^2 S^2 c_{SS} = r_0 c + \frac{\mu - r_0}{\sigma} \sigma S c_S.
$$

Rearranging results in

$$
c_t + r_0 S c_S + \frac{1}{2}\sigma^2 S^2 c_{SS} = r_0 c, \tag{6.6}
$$

which is the celebrated Black–Scholes partial differential equation for option pricing!

At this point, we have derived an absence of arbitrage condition in the form of the partial differential equation (6.6). This equation applies to a generic derivative security that can be modeled as a twice continuously differentiable function of the stock price and time, $c(S, t)$. The final step is to consider the payoff characteristics of a specific derivative security and actually solve the Black–Scholes equation under those conditions. The solution for European call and put options is given next.

6.1.1   Solution for European Call and Put Options

A European call option with strike price K and expiration T gives the holder the right, but not the obligation, to purchase a share of the stock S at time T for the price K. As explained in Example 5.8 of Chapter 5, this gives the European call option the payoff function:

$$c(S(T), T) = \max\{S(T) - K, 0\}.$$

The solution to the Black–Scholes partial differential equation subject to this payoff function as a boundary condition is

$$c_{BS}(S, t) = S\Phi(d_1) - Ke^{-r_0(T-t)}\Phi(d_2) \tag{6.7}$$

where

$$d_1 = \frac{\ln(S/K) + (r_0 + \frac{1}{2}\sigma^2)(T - t)}{\sigma\sqrt{T - t}}, \tag{6.8}$$

$$d_2 = d_1 - \sigma\sqrt{T - t}, \tag{6.9}$$

and $\Phi(\cdot)$ is the cumulative distribution function of the standard normal random variable. This is the famous Black–Scholes formula for a European call option!

I won't provide a derivation of this solution here, but rather suggest that you may verify for yourself that it indeed is the solution by substituting it into the Black–Scholes partial differential equation (6.6). However, I will note that this solution and minor variants of it will take us incredibly far. The reason for this is that so much intuition has been built up around this classic Black–Scholes formula that many other models are made to "look" like it so that similar intuition will be applicable. Thus, I believe you will be rewarded handsomely for spending a little time to become familiar with this formula.

As a curiosity, note that the Black–Scholes formula does not depend on the drift rate μ of the stock. This rather surprising result, and more general ideas about what features of our models don't affect pricing, will lead us down the path to risk neutral pricing in Chapter 10. For now, simply recognize that this lack of dependence is an advantageous feature of the Black–Scholes formula, since the drift rate μ of a stock is extremely hard to estimate accurately.

A closed form solution also exists for European put options. Recall that a European put option with strike price K and expiration T gives the holder the right, but not the obligation, to *sell* a share of the stock S at time T for the price K. If we denote the value of the put option by $p(S,t)$, then as detailed in Example 5.8 of Chapter 5, the payoff function is

$$p(S(T),T) = \max\{K - S(T), 0\}.$$

This put must also satisfy the Black–Scholes partial differential equation (6.6). Once again, a closed form solution is available and given by

$$p_{BS}(S,t) = Ke^{-r_0(T-t)}\Phi(-d_2) - S\Phi(-d_1) \tag{6.10}$$

where d_1 and d_2 have already been defined in equations (6.8) and (6.9). This is the Black–Scholes formula for a European put option.

Let's work through two examples involving the call and put formulas.

Example 6.1 (Black–Scholes Call and Put Formula)
Let $S(0) = \$10.00$ be the current price of a non-dividend paying stock, and assume that the stock has drift $\mu = 10\%$ and volatility $\sigma = 30\%$.

Consider European call and put options, both with strike price $K = \$10.00$ and time to expiration of a quarter of a year, $T = 0.25$. Let the continuously compounded risk-free rate of interest be $r_0 = 5\%$.

To calculate the Black–Scholes price of these options, we first compute d_1 and d_2 given in equations (6.8) and (6.9),

$$d_1 = \frac{\ln(10.00/10.00) + (0.05 + \frac{1}{2}(0.25)^2)(0.25)}{(0.3)\sqrt{0.25}} = 0.15833,$$

$$d_2 = d_1 - (0.3)\sqrt{0.25} = 0.00833.$$

We then substitute these numbers into the Black–Scholes formulas for the call and put in equations (6.7) and (6.10) to obtain the values,

$$c_{BS} = 10.00\Phi(0.15833) - 10.00e^{-(0.05)(0.25)}\Phi(0.00833)) = 0.6583,$$

$$p_{BS} = 10.00e^{-(0.05)(0.25)}\Phi(-0.00833) - 10.00\Phi(-0.15833) = 0.5341.$$

Thus, we find that the call option is slightly more valuable than the put option.

Example 6.2 (Graphical Representation of Black–Scholes Formula)
A convenient way to understand the Black–Scholes call and put formulas is to display them graphically, as a function of the initial stock price S on a plot that also includes the payoff functions of the call and put. Using the parameter values $T = 0.25$, $K = \$10.00$, $\sigma = 0.3$, and $r_0 = 0.05$, the Black–Scholes value of a European call and put option as a function of S is shown as the solid line in Figure 6.1, where the upper plot is for the call option and the lower plot displays the put option. For comparison, the payoff functions of the call and put, which

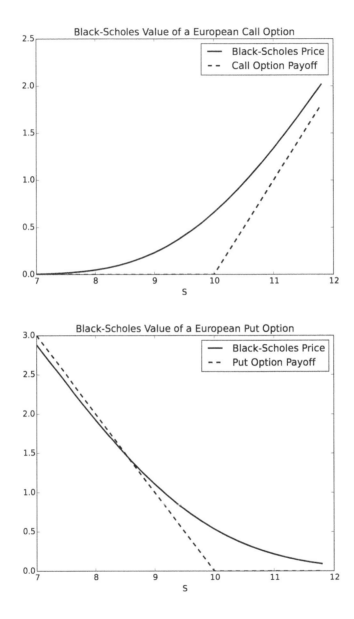

Figure 6.1   Plot of the payoff and Black–Scholes price of call (top) and put (bottom) option as a function of the current stock price for the parameter values $K = \$10.00$, $T = 0.25$, $\sigma = 0.3$, and $r_0 = 0.05$.

are $\max\{S - K, 0\} = \max\{S - \$10, 0\}$ and $\max\{K - S, 0\} = \max\{\$10 - S, 0\}$, respectively, are provided as the dotted lines.

From the upper plot for the call option, we see that the value of the call approaches zero as the option becomes increasingly out-of-the-money (refer to Example 5.8 of Chapter 5 for "moneyness" terminology). That is, for a stock price S that is far below the strike price $K = \$10.00$, the value of the option is close to zero. On the other hand, for values of the stock price S greater than the strike price when the option is in-the-money, the value of the call grows with the increase in the stock price. This should correspond to your intuition for a call option, since the value of being able to purchase the stock at a price of $K = \$10.00$ should grow with the increasing value of the stock S.

The intuition is similar, just reversed, for the put option. The bottom plot shows that the value of the put option increases as the price S decreases, and decreases as the price S increases.

6.1.2   Implied Volatility

When we price derivative securities, we price them in terms of the underlying variables which we think of as being "known." Moreover, by extension, there is also the implicit assumption that the factor models of the underlying variables are also known.

In the case of the Black–Scholes model, the volatility of the stock σ is a parameter in the solution. In fact, if we explicitly write the Black–Scholes formula for a call and put option as a function of all the parameters that appear, we would write $c_{BS}(S, t, K, T, r_0, \sigma)$ and $p_{BS}(S, t, K, T, r_0, \sigma)$, respectively. That is, the formulas depend on the stock price S, the current time t, the strike price K, the expiration time T, and risk-free rate r_0, and finally the option volatility σ. Of these, all are known and directly observable in the market except for the volatility σ. To emphasize that the Black–Scholes formula requires all these parameters as inputs, we will sometimes explicitly write out all the arguments as

$$c_{BS}(S, t) = c_{BS}(S, t, K, T, r_0, \sigma), \quad p_{BS}(S, t) = p_{BS}(S, t, K, T, r_0, \sigma) \qquad (6.11)$$

where $c_{BS}(S, t)$ is the Black–Scholes call formula in (6.7), and $p_{BS}(S, t)$ is the Black–Scholes put formula in (6.10).

Because volatility is not directly observable, we would have to *estimate* the value of the volatility of the stock to apply the Black–Scholes formula. However, in practice, it is common to actually use the market prices of options, which are *known and observable* in the market, to go "backward" to obtain the value for the volatility that is consistent with the market. When the volatility of the stock is obtained in this manner, it is called the *implied volatility*, which we will denote by σ_i. To illustrate this idea, consider the following example.

Example 6.3 (Implied Volatility)
Consider a stock with the current price of $S(0) = \$12$. Furthermore, assume that there is a European call option with strike price $K = \$10$, time to expiration

$T = 0.25$, and current market price $c_m = \$2.159$. The risk-free rate of interest is $r_0 = 5\%$.

The implied volatility, σ_i, of the stock is found by solving for the value of σ that makes the Black–Scholes formula equal to the market price of the option. That is, let $c_{BS}(S, t, K, T, r, \sigma)$ be the Black–Scholes formula for the price of a European call. We must solve the equation

$$c_m = 2.159 = c_{BS}(12, 0, 10, 0.25, 0.05, \sigma_i) = c_{BS}(S, t, K, T, r_0, \sigma_i)$$

for σ_i. In this specific case, you may verify that the answer is $\sigma_i = 25\%$. That is, the implied volatility of the stock is $\sigma_i = 25\%$ since

$$c_{BS}(12, 0, 10, 0.25, 0.05, 0.25) = 2.159.$$

While the Black–Scholes equation and corresponding pricing formulas for European calls and puts are fundamental to the field of derivative pricing, numerous extensions and variations have been developed since it was first published in 1973. In the rest of this chapter, we consider a wide array of such other models and derive their corresponding absence of arbitrage conditions in the context of the factor model framework.

6.2   CONTINUOUS DIVIDENDS

Merton [39] extended the Black–Scholes approach so that it would apply to stocks that pay a continuous dividend. In this case, we must be careful about our classification of variables and make sure that we know exactly what is tradable.

Step 1: Model and Classify Variables
Consider a market with a bond B paying a risk-free rate of r_0 and a stock with price per share $S(t)$ that pays a continuous dividend at a rate of q. What this means is that from time period t to $t + dt$, for each share of the stock we hold, we will receive a dividend in the amount of $qdtS(t)$.

One of the most convenient methods of modeling this is to assume that any dividend is reinvested in the stock. That is, we use the dividend amount $qdtS(t)$ to purchase more shares of the stock. Since the stock price is $S(t)$, an amount $qdtS(t)$ will allow us to purchase qdt additional shares of the stock over the period dt.

More generally, if we let $n(t)$ denote the number of shares held of the stock at time t, then the continuous dividend rate of q means that $n(t)$ grows over the next time increment dt according to the equation

$$dn = qndt. \tag{6.12}$$

Additionally, we will model the price per share $S(t)$ as a standard geometric Brownian motion,

$$dS = \mu S dt + \sigma S dz. \tag{6.13}$$

Now, when classifying variables, it is important to note that the stock and its dividend stream *together* constitute the tradable. That is, when we purchase the stock, we are also purchasing its dividend stream. Hence we must consider them together as a tradable asset. Thus, we need to derive a factor model for the change in the value of the shares that we hold.

To do this, let's denote the value of $n(t)$ shares along with the associated dividend stream as $v(t)$. Initially, we have $v(t) = n(t)S(t)$. To obtain a factor model for this, we apply Ito's product rule, noting that n is governed by equation (6.12) and S by equation (6.13):

$$
\begin{aligned}
dv(t) \quad &- \quad d(n(t)S(t)) \\
&= \quad S(t)dn(t) + n(t)dS(t) + dn(t)dS(t) \\
&= \quad qn(t)S(t)dt + (\mu n(t)S(t)dt + \sigma n(t)S(t)dz) \\
&= \quad qv(t)dt + \mu v(t)dt + \sigma v(t)dz \\
&= \quad (\mu + q)v(t)dt + \sigma v(t)dz.
\end{aligned}
$$

Thus, $v(t)$ represents the tradable quantity and is governed by the stochastic differential equation

$$ dv = (\mu + q)v(t)dt + \sigma v(t)dz. \tag{6.14} $$

Finally, we consider a derivative security on the stock whose payoff depends on the price of the stock alone, and not on the dividend stream. For example, in the case of a European call option, the payoff of the option depends on the price per share $S(T)$ at expiration. Therefore, we assume that the derivative can be written as $c(S(t), t)$. An application of Ito's lemma leads to

$$ dc = (c_t + \mu S c_S + \frac{1}{2}\sigma^2 S^2 c_{SS})dt + \sigma S c_S dz. \tag{6.15} $$

With this modeling complete, we classify the variables as

Tradables	Underlying Variables	Factors
B, v, c	r_0, S	dt, dz.

Step 2: Construct a Tradable Table

To create a tradable table, we need models for the tradables. The risk-free bond B follows our standard model under continuous compounding of $dB = r_0 B dt$. We also have a model for the value of the stock v given in equation (6.14) and for the derivative in equation (6.15). These allow us to write the tradable table as

$$
\begin{array}{cc}
\text{Prices} & \text{Value Change Factor Models} \\
\end{array}
$$

$$
\begin{bmatrix} B \\ v \\ \hline c \end{bmatrix} \quad d \begin{bmatrix} B \\ v \\ \hline c \end{bmatrix} = \begin{bmatrix} r_0 B \\ (\mu + q)v \\ \hline c_t + \mu S c_S + \frac{1}{2}\sigma^2 S^2 c_{SS} \end{bmatrix} dt + \begin{bmatrix} 0 \\ \sigma v \\ \hline \sigma S c_S \end{bmatrix} dz.
$$

Note that the stock price $S(t)$ alone does not appear in the tradable table since it is not tradable. As a consequence, it is not directly constrained by the absence of arbitrage conditions.

Step 3: Apply the Price APT

The final step is to solve the Price APT equation $\mathcal{A} = \mathcal{P}\lambda_0 + \mathcal{B}\lambda_1$, which is

$$
\begin{bmatrix} r_0 B \\ (\mu + q)v \\ -- \\ c_t + \mu S c_S + \frac{1}{2}\sigma^2 S^2 c_{SS} \end{bmatrix} = \begin{bmatrix} B \\ v \\ -- \\ c \end{bmatrix} \lambda_0 + \begin{bmatrix} 0 \\ \sigma v \\ -- \\ \sigma S c_S \end{bmatrix} \lambda_1 \qquad (6.16)
$$

As usual, we solve first using the marketed tradables above the horizontal dashes, giving λ_0 and λ_1 as

$$
\lambda_0 = r_0 \qquad \lambda_1 = \frac{\mu + q - r_0}{\sigma}, \qquad (6.17)
$$

followed by substituting these quantities into the last equation for the derivative,

$$
c_t + \mu S c_S + \frac{1}{2}\sigma^2 S^2 c_{SS} = r_0 c + \frac{\mu + q - r_0}{\sigma}\sigma S c_S. \qquad (6.18)
$$

Finally, rearranging leads to the absence of arbitrage partial differential equation,

$$
c_t + (r_0 - q)S c_S + \frac{1}{2}\sigma^2 S^2 c_{SS} = r_0 c. \qquad (6.19)
$$

Comparing this to the Black–Scholes equation obtained in (6.6) reveals only slight differences. In fact, it is common to call this the Black–Scholes–Merton equation. It is not surprising that a closed form solution for European calls and puts also exists in this case.

6.2.1 European Calls and Puts under Continuous Dividends

Once again, we can consider the value of European call and put options, but this time on a stock that pays a continuous dividend. The closed form solutions are

$$
\begin{aligned}
c(S,t) &= S e^{-q(T-t)}\Phi(d_1) - K e^{-r_0(T-t)}\Phi(d_2), \\
p(S,t) &= K e^{-r_0(T-t)}\Phi(-d_2) - S e^{-q(T-t)}\Phi(-d_1)
\end{aligned}
$$

where

$$
\begin{aligned}
d_1 &= \frac{\ln(S/K) + (r_0 - q + \frac{1}{2}\sigma^2)(T-t)}{\sigma\sqrt{T-t}}, \qquad (6.20) \\
d_2 &= d_1 - \sigma\sqrt{T-t}. \qquad (6.21)
\end{aligned}
$$

Note that these formulas are very similar to the formulas for European call and put options on a non-dividend paying stock; the only difference is that q shows up in a couple places. Let's work an example that illustrates the role that the dividend rate q plays in the valuation of calls and puts.

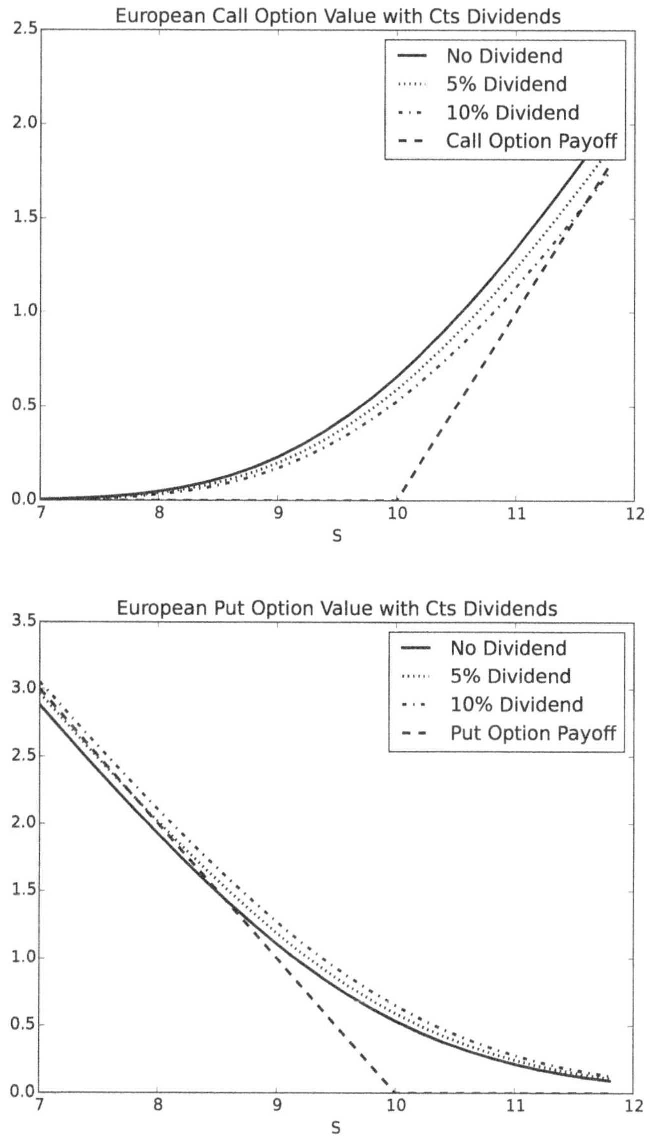

Figure 6.2 Value of call (top) and put (bottom) options as a function of the initial stock price S for the cases of no dividends and dividend rates of $q = 5\%$ and $q = 10\%$. Note that an increased dividend rate causes call options to be less valuable and put options to be more valuable.

Example 6.4 (Calls and Puts under Continuous Dividends)
Consider European call and put options, both with strike price $K = \$10$ and time to expiration $T = 0.25$, on a stock that pays a continuous dividend. Let the volatility of the stock be $\sigma = 0.3$ and assume the risk-free rate is $r_0 = 5\%$. In this example,

we will vary the continuous dividend rate q that the stock pays and observe its effect on the value of the call and put.

Figure 6.2 shows plots of the price of the call (top) and put (bottom) as functions of the current price of the stock S and for the three dividend rates of $q = 0$, $q = 5\%$ and $q = 10\%$. Observe in the plots that the value of the call option *decreases* as the dividend rate increases, while the value of the put option *increases* as the dividend increases. Why is this so?

In the case of a call option, if the stock pays a large dividend, it essentially "bleeds" away at the price per share $S(t)$ over time since the value of the company that the share represents decreases when it pays dividends. This means that the larger the dividend, the more likely the share price at expiration will not exceed the strike price, thus making the call option less valuable.

The reasoning for the put option is similar. In this case, the put gives the holder the right to sell at a fixed price. If the share price is smaller at expiration, which will be the effect of paying a large dividend rate, then this benefits the put option holder, thus making puts more valuable.

6.2.2 What Pays a Continuous Dividend?

At first thought, a continuous model for a dividend does not seem too realistic. However, a moment of thought more and we realize that it is a decent approximation for a number of financial situations. For instance, a stock index comprised of many stocks that pay dividends at different times can be approximated as a continuous dividend. So can a commodity with a convenience yield. Moreover, foreign currencies invested in a money market account earning some continuously compounded rate of interest are essentially securities that earn a continuous dividend. So, the point is, don't let the notation of calling S a "stock" limit your thinking about how these models can be applied.

6.3 CASH DIVIDENDS

Most individual stocks pay a prespecified dividend at a prespecified time. We often call this either a *cash* or *discrete* or *lump* dividend. To obtain the equation satisfied by an option on a stock that pays a cash dividend prior to expiration, we will once again walk through our three step procedure.

Step 1: Model and Classify Variables
We always start with a tradable bond following $dB = r_0 B dt$. Let's assume that we know our end goal is to price a European call or put that expires at time T. Moreover, assume that the stock on which the option is written pays a known dividend amount D_τ at the time τ which is prior to expiration. That is, $\tau < T$.

Here is the simplest way to model this situation. Since we know that we are going to be paid a dividend at time τ in an amount of D_τ (let's assume there is only one dividend payment prior to expiration T), then we can essentially model the value of a share of stock as the sum of two parts. The first part is the value of the known dividend that we will receive at time τ. Since the dividend is known and deterministic, we simply value it by discounting it to the current time t at the risk-free rate r_0. Thus, the value of the dividend at time $t \leq \tau$ is $D_\tau e^{-r_0(\tau-t)}$. The second part of the value of a share accounts for the "rest" of the company, which we denote by $S(t)$ and assume follows a geometric Brownian motion,

$$dS = \mu S dt + \sigma S dz.$$

Thus, the total value $v(t)$ of a share prior to the dividend date τ is just the sum of the two parts,

$$v(t) = S(t) + D_\tau e^{-r_0(\tau-t)}, \quad t \leq \tau. \tag{6.22}$$

After the dividend has been paid (i.e., for $t > \tau$), the value of a share is simply

$$v(t) = S(t), \quad t > \tau.$$

Therefore, to obtain the factor model for the value of a share, $v(t)$, we break it up into the two cases above. For $t \leq \tau$, taking the differential of $v(t)$ in equation (6.22) leads to

$$dv = dS + r_0 D_\tau e^{-r_0(\tau-t)} dt = \left(\mu S + r_0 D_\tau e^{-r_0(\tau-t)} \right) dt + \sigma S dz. \tag{6.23}$$

After the dividend is paid, $t > \tau$, we have

$$dv = dS = \mu v dt + \sigma v dz. \tag{6.24}$$

These two equations, (6.23) and (6.24), provide us with our factor model for $v(t)$.

Finally, we consider the modeling of the derivative c. Since we have in mind European call and put options with expiration T, their payoff will only depend on the value $S(T)$ and not on the dividend. Thus, we assume that the derivative can be represented as functions of $S(t)$ and t as in $c(S(t), t)$. Ito's lemma gives,

$$dc = \left(c_t + \mu S c_S + \frac{1}{2} \sigma^2 S^2 c_{SS} \right) dt + \sigma S c_S dz.$$

With this understanding, we classify the variables as

Tradables	Underlying Variables	Factors
B, v, c	r_0, S	dt, dz.

Step 2: Construct a Tradable Table

In this case, we will need two tradable tables, one for $t \leq \tau$ and one for $t > \tau$. For $t \leq \tau$, the tradable table is

Prices   Value Change   Factor Models

$$\begin{bmatrix} B \\ v \\ -- \\ c \end{bmatrix} \quad d \begin{bmatrix} B \\ v \\ -- \\ c \end{bmatrix} = \begin{bmatrix} r_0 B \\ \mu S + r_0 D_\tau e^{-r_0(\tau-t)} \\ -- \\ c_t + \mu S c_S + \frac{1}{2}\sigma^2 S^2 c_{SS} \end{bmatrix} dt + \begin{bmatrix} 0 \\ \sigma S \\ -- \\ \sigma S c_S \end{bmatrix} dz$$

where $v(t) = S(t) + D_\tau e^{-r_0(\tau-t)}$.

The tradable table for $t > \tau$ is identical to the tradable table in the standard Black–Scholes case. It will result in the same market prices of risk and Black–Scholes equation (6.6). So, going forward, let's focus on the tradable table above and see where it leads.

Step 3: Apply the Price APT

Let's take the tradable table above and apply the Price APT. This leads to the equations,

$$
\begin{bmatrix}
r_0 B \\
\mu S + r_0 D_\tau e^{-r_0(\tau-t)} \\
-- \\
c_t + \mu S c_S + \frac{1}{2}\sigma^2 S^2 c_{SS}
\end{bmatrix}
=
\begin{bmatrix}
B \\
v \\
-- \\
c
\end{bmatrix}
\lambda_0 +
\begin{bmatrix}
0 \\
\sigma S \\
-- \\
\sigma S c_S
\end{bmatrix}
\lambda_1.
$$

Solving the first equation gives $\lambda_0 = r_0$, as usual. The second equation is then

$$
\mu S(t) + r_0 D_\tau e^{-r_0(\tau-t)} = v(t)\lambda_0 + \sigma S(t)\lambda_1.
$$

Upon substitution of $v(t) = S(t) + D_\tau e^{-r_0(\tau-t)}$ and $\lambda_0 = r_0$, it reduces to

$$
\mu S(t) = r_0 S(t) + \sigma S(t)\lambda_1,
$$

which leads to $\lambda_1 = \frac{\mu - r_0}{\sigma}$. This is the same result that we have in the Black–Scholes no-dividend case. Substituting into the last equation leads to the standard Black–Scholes partial differential equation,

$$
c_t + r_0 S c_S + \frac{1}{2}\sigma^2 S^2 c_{SS} = r_0 c.
$$

Since the tradable table for $t > \tau$ also leads to this same equation, we see that the Black–Scholes equation governs the price of a European call or put both before and after the dividend is paid. However, note that prior to the payment of the dividend, the value of a share observed in the market is $v(t) = S(t) + D_\tau e^{-r_0(\tau-t)}$, and thus to determine the value of $S(t)$, we need to take $v(t)$ and subtract the discounted value of the dividend $D_\tau e^{-r_0(\tau-t)}$.

This leads us to a Black–Scholes based closed form solution for European calls and puts on a stock paying a known cash dividend, as described next.

6.3.1 European Calls and Puts under Cash Dividends

To price European calls and puts under cash dividends at time t, we note that one may simply discount the dividend payment, which occurs at time τ, back to time t, and assume that the stock started at the initial value $S(t) = v(t) - De^{-r_0(\tau-t)}$ rather than at $v(t)$. This is the same as if the company simply decided to pre-pay us the dividend at time t instead of at the dividend date τ. To account for the time difference of the pre-payment, we only receive the discounted value of the dividend at time t which is $De^{-r_0(\tau-t)}$.

In this case, we can directly apply the Black–Scholes formulas for calls and puts in equations (6.7) and (6.10), but using $S(t) = v(t) - De^{-r_0(\tau-t)}$ for the initial stock price.

Example 6.5 (Cash Dividend)
Consider a European call option with strike price $K = \$10$ and expiration $T = 0.25$ on a stock with current price 12 and volatility $\sigma = 0.3$. In the model developed in this section, we consider the value of a share at time $t = 0$ to be $v(0) = \$12$. Assume the stock is set to pay a dividend of 1 with a dividend date occurring at time $\tau = 0.1$. If the risk-free rate is $r_0 = 0.05$, then the value of this option can be computed using the Black–Scholes formula (6.7) as follows.

First, calculate the initial value of the stock minus the discounted value of the dividend to obtain $S(0)$:

$$S(0) = v(0) - De^{-r_0\tau} = 12 - 1e^{-0.05(0.1)} = 11.005.$$

Next, use $S(0)$ as the initial price of the stock and substitute into the Black–Scholes formula. That is, taking $S = 11.005$, $t = 0$, $\sigma = 0.3$, $K = 10$, $T = 0.25$, and $r_0 = 0.05$ in the Black–Scholes formula gives

$$c = 1.344$$

which is the price of the European call option.

6.4 POISSON PROCESSES

Cox and Ross [13] created a highly stylized model in which they replaced the Brownian factor by a Poisson process. This leads to a Poisson version of the Black–Scholes formula. We will derive the absence of arbitrage equation governing this situation.

Step 1: Model and Classify Variables
This setup involves a bond earning a risk-free rate and a non-dividend paying stock that follows a geometric Poisson motion,

$$dB = r_0 B dt \qquad (6.25)$$
$$dS = \mu S^- dt + (k-1)S^- d\pi(\alpha) \qquad (6.26)$$

where α is the intensity of the Poisson process $\pi(t; \alpha)$. Geometric Poisson motion of this form was considered in Chapter 3, Section 3.2.

Recall that the simple binary approximation to the Poisson process presented in Chapter 1 is that $d\pi(t; \alpha)$ jumps to a value of 1 with probability αdt and stays at 0 with probability $1 - \alpha dt$. In this case, viewing the geometric Poisson motion of equation (6.26) in that context, the Poisson process induces a model of the stock so that with probability αdt the stock jumps from S^- to kS^-; otherwise, with probability $1 - \alpha dt$, it just drifts at a constant rate of μ.

Now, consider a derivative on the stock that can be represented as $c(S, t)$ and is continuously differentiable in both its arguments. By the version of Ito's lemma for Poisson processes, we have

$$dc = (c_t + \mu S^- c_S)dt + (c(kS^-) - c(S^-))d\pi(\alpha). \tag{6.27}$$

This is a case where the random factor $d\pi(\alpha)$ does not have zero mean. I would like to write my factor equations such that the factors are pure risk and don't have any expected drift. Hence, I will compensate the factor to give it zero drift.

The easiest way to do this is to subtract and then add back in the expected drift of the factor. That is, we make the adjustment

$$d\pi(\alpha) = (d\pi(\alpha) - \alpha dt) + \alpha dt = d\pi^\alpha + \alpha dt$$

where

$$d\pi^\alpha = (d\pi(\alpha) - \alpha dt)$$

is the new compensated factor. Using this substitution in the equations for the stock and derivative results in

$$
\begin{aligned}
dS &= (\mu S^- + \alpha(k-1)S^-)dt + (k-1)S^- d\pi^\alpha, \\
dc &= (c_t + \mu S^- c_S + \alpha(c(kS^-) - c(S^-)))dt + (c(kS^-) - c(S^-))d\pi^\alpha.
\end{aligned}
$$

Finally, the classification of variables is straightforward in this case,

Tradables	Underlying Variables	Factors
B, S, c	r_0, S	dt, $d\pi$.

Step 2: Construct a Tradable Table
Using the modeling above under the compensated factor $d\pi^\alpha$, we construct the tradable table as

Prices   Value Change Factor Models

$$
\begin{bmatrix} B \\ S^- \\ -- \\ c \end{bmatrix}
\quad d\begin{bmatrix} B \\ S \\ -- \\ c \end{bmatrix}
= \begin{bmatrix} r_0 B \\ \mu S^- + \alpha(k-1)S^- \\ -- \\ c_t + \mu S^- c_S + \alpha(c(kS^-) - c(S^-)) \end{bmatrix} dt
+ \begin{bmatrix} 0 \\ (k-1)S^- \\ -- \\ c(kS^-) - c(S^-) \end{bmatrix} d\pi^\alpha.
$$

Step 3: Apply the Price APT
With the tradable table given above, we can move on to Step 3, which is to apply the Price APT equation $\mathcal{A} = \mathcal{P}\lambda_0 + \mathcal{B}\lambda_1$ for the absence of arbitrage condition,

$$
\begin{bmatrix} r_0 B \\ \mu S^- + \alpha(k-1)S^- \\ -- \\ c_t + \mu S^- c_S + \alpha(c(kS^-) - c(S^-)) \end{bmatrix}
= \begin{bmatrix} B \\ S^- \\ -- \\ c \end{bmatrix} \lambda_0
+ \begin{bmatrix} 0 \\ (k-1)S^- \\ -- \\ c(kS^-) - c(S^-) \end{bmatrix} \lambda_1.
$$

Solving for λ_0 and λ_1 gives

$$\lambda_0 = r_0, \qquad \lambda_1 = \frac{\mu - r_0}{k - 1} + \alpha.$$

Substituting λ_0 and λ_1 into the last equation for the derivative yields

$$c_t + \mu S^- c_S + \alpha(c(kS^-) - c(S^-)) = r_0 c + \left(\frac{\mu - r_0}{k - 1} + \alpha\right)(c(kS^-) - c(S^-)).$$

Finally, rearranging gives the absence of arbitrage equation,

$$(k - 1)\left(c_t + \mu S^- c_S - r_0 c\right) = (\mu - r_0)(c(kS^-) - c(S^-)). \tag{6.28}$$

6.4.1 Closed Form Solution for European Call Option

One can actually find a closed form solution for a European call option with strike K and expiration T under this model. For a call option with strike K and expiration T, the formula is

$$c(S, t) = S\Psi(x, y) - Ke^{-r_0(T-t)}\Psi(x, y/k) \tag{6.29}$$

where

$$\Psi(\alpha, \beta) = \sum_{i=\alpha}^{\infty} \frac{e^{-\beta}\beta^i}{i!}, \quad y = \frac{(r_0 - \mu)(T - t)k}{k - 1}, \tag{6.30}$$

and x is the smallest non-negative integer greater than $\frac{ln(K/S) - \mu(T-t)}{ln(k)}$.

Note that this solution has a structure that is very similar to the Black–Scholes formula In fact, one can interpret it as the Black–Scholes formula with a Poisson random variable replacing the Gaussian random variable.

A closed form solution for a European put option exists as well, but is not given here. The model is not terribly practical due to the rather unrealistic nature of the stock price movement that it assumes, but its derivation is nevertheless an instructive exercise.

6.5 OPTIONS ON FUTURES

Pricing an option on a futures contract requires some careful thought. If you recall, futures posed a special case in our Return APT equations of Chapter 4, but were handled in a more direct manner using the Price APT. In this section, we will derive the equation for a derivative depending on a futures price. This was done by Black in [2].

Step 1: Model and Classify Variables

Assume there exists a bond B and a futures contract whose value I will denote by $\mathcal{F}$, with corresponding futures price f given as

$$dB = r_0 B dt, \tag{6.31}$$
$$df = \mu f dt + \sigma f dz. \tag{6.32}$$

Recall that the futures price f refers to the price that has been agreed upon now, but will be paid at the expiration of the futures contract in exchange for some asset. The mark to market mechanism of futures contracts is one of their distinguishing features, especially as compared to forward contracts. If you don't recall how that works, please refer back to the discussion in Chapter 4, Section 4.4.2.

Here we consider a derivative on the futures price. Generically we call this derivative c and assume that its price process depends on the futures price and time $c(f, t)$, and is twice continuously differentiable. An application of Ito's lemma results in the stochastic differential equation,

$$dc = (c_t + \mu f c_f + \frac{1}{2}\sigma^2 f^2 c_{ff})dt + \sigma f c_f dz. \tag{6.33}$$

Finally, our classification of quantities is:

Tradables	Underlying Variables	Factors
B, $\mathcal{F} = 0$, c	r_0, f	dt, dz.

Note how we have made a distinction between the value of the futures contract $\mathcal{F}$, which is tradable, and the futures price related to the contract f, which serves as an underlying variable. Recalling our discussion of futures contracts from Chapter 4, Section 4.4.2, the exact mechanics of the futures contract will play into our tradable table. Specifically, futures contracts undergo the mark to market mechanism in which the value of a contract $\mathcal{F}$ is reset to zero for the start of each day, and a cash flow in the amount of df occurs at the end of each day. That is, due to the mark to market mechanism, we model the price of a futures contract as $\mathcal{F} = 0$, and the cash flow resulting from holding a futures contract over a time period dt as df.

Step 2: Construct a Tradable Table
Taking the above discussion of the unique characteristics of futures contracts into consideration, the tradable table is

$$
\begin{array}{cc}
\text{Prices} & \text{Value Change} \quad \text{Factor Models} \\
\end{array}
$$

$$
\begin{bmatrix} B \\ \mathcal{F} = 0 \\ -- \\ c \end{bmatrix} \quad
d\begin{bmatrix} B \\ f \\ -- \\ c \end{bmatrix} =
\begin{bmatrix} r_0 B \\ \mu f \\ -- \\ c_t + \mu f c_f + \frac{1}{2}\sigma^2 f^2 c_{ff} \end{bmatrix} dt +
\begin{bmatrix} 0 \\ \sigma f \\ -- \\ \sigma f c_f \end{bmatrix} dz.
$$

To reiterate, the key point is that the mark to market mechanism always sets the price of a futures contract to zero, while the cash flow from the contract over period dt is given by the change in the futures price df. Please take note that this is reflected in the above tradable table.

Step 3: Apply the Price APT

With the above tradable table, we can proceed in a straightforward manner to apply the Price APT equation $\mathcal{A} = \mathcal{P}\lambda_0 + \mathcal{B}\lambda_1$, leading to

$$
\begin{bmatrix}
r_0 B \\
\mu f \\
-- \\
c_t + \mu f c_f + \frac{1}{2}\sigma^2 f^2 c_{ff}
\end{bmatrix}
=
\begin{bmatrix}
B \\
0 \\
-- \\
c
\end{bmatrix}
\lambda_0 +
\begin{bmatrix}
0 \\
\sigma f \\
-- \\
\sigma f c_f
\end{bmatrix}
\lambda_1 .
$$

Solving for λ_0 and λ_1 gives

$$
\lambda_0 = r_0 \qquad \lambda_1 = \frac{\mu}{\sigma}.
$$

Substituting λ_0 and λ_1 into the last equation yields

$$
c_t + \mu f c_f + \frac{1}{2}\sigma^2 f^2 c_{ff} = r_0 c + \frac{\mu}{\sigma}\sigma f c_f,
$$

and rearranging results in the partial differential equation:

$$
c_t + \frac{1}{2}\sigma^2 f^2 c_{ff} = r_0 c. \tag{6.34}
$$

6.5.1 Solution for European Call and Put Options

The partial differential equation in (6.34) looks just like the Black–Scholes–Merton equation on a stock paying a continuous dividend (6.19) with the dividend rate set equal to the risk-free rate, $q = r_0$. Therefore, we can utilize that formula to obtain the prices of European call and put options on futures as

$$
\begin{aligned}
c(f,t) &= e^{-r_0(T-t)}\left(f\Phi(d_1) - K\Phi(d_2)\right), \tag{6.35} \\
p(f,t) &= e^{-r_0(T-t)}\left(K\Phi(-d_2) - f\Phi(-d_1)\right) \tag{6.36}
\end{aligned}
$$

with

$$
\begin{aligned}
d_1 &= \frac{\ln(f/K) + (\frac{1}{2}\sigma^2)(T-t)}{\sigma\sqrt{T-t}}, \tag{6.37} \\
d_2 &= d_1 - \sigma\sqrt{T-t}. \tag{6.38}
\end{aligned}
$$

These formulas are commonly referred to as Black's model. The numerical example next shows how these formulas are applied

Example 6.6 (Calls and Puts on Futures)
Consider call and put options on a futures contract where the current futures price is $f = \$100$. Moreover, assume that the volatility of the futures price is $\sigma = 0.4$ and the risk-free rate is $r_0 = 0.05$.

If the strike price and expiration are $K = \$105$ and $T = 0.1$, then the values of call and put options are given by equations (6.35) and (6.36), leading to

$$c = 3.033, \quad p = 8.008.$$

6.6 JUMP DIFFUSION MODEL

A jump diffusion model in the context of option pricing was first considered by Merton in [40]. This model is nice because it is related to many other models in equity and interest rate derivatives. The model includes a risk-free bond and an underlying asset that has a diffusion portion and a jump portion. Jump diffusion stochastic differential equations were considered in Chapter 3, Section 3.3. Quite conveniently, this model admits a closed form solution for the European calls and puts when the jumps are lognormal, as Merton computed in his original paper.

We will also find that this model creates a problem for our factor approach. However, it is possible to bypass it with some sleight of hand. Merton addressed the issue that arises using a similar technique, so we will follow his lead.

Step 1: Model and Classify Variables
The basic assets are a bond B and a non-dividend paying stock S following

$$dB = r_0 B dt, \tag{6.39}$$
$$dS = \mu S^- dt + \sigma S^- dz + S^-(Y - 1)d\pi(\alpha). \tag{6.40}$$

The stock S is driven by the Brownian factor dz and also a compound Poisson component $(Y-1)d\pi(\alpha)$ where the variable that controls the jump size Y is random. Recall that the parameter α is the intensity of the Poisson process, and controls the likelihood of a jump. In particular, the expected number of jumps in a year is α.

We will see that this model is awkward to fit into our linear factor model framework because of the $(Y - 1)d\pi$ term. It would be convenient if we could just view $d\pi$ as the factor, but the $(Y - 1)$ portion also contains randomness. Thus we will be forced to treat the entire term $(Y - 1)d\pi$ as a factor. However, we will see that this is problematic because we will lose the linear structure of our factor model when we consider the derivative security.

Before we get to that point, let's deal with another issue. The term $(Y - 1)d\pi$ may have non-zero expectation. I would like to view this term as a pure risk factor, with zero expectation. To do so, I need to compensate this term by subtracting its mean $\alpha \mathbb{E}[Y - 1]dt$ and adding it back into the drift portion of the stochastic differential equation. That is, let's rewrite the equation for dS as

$$dS = (\mu + \alpha \mathbb{E}[Y - 1])S^- dt + \sigma S^- dz + S^-((Y - 1)d\pi(\alpha) - \alpha \mathbb{E}[Y - 1]dt).$$

By writing the stochastic differential equation this way, I can now view the entire quantity $((Y - 1)d\pi(\alpha) - \alpha \mathbb{E}[Y - 1]dt)$ as a factor with zero mean. This appears to be just a cosmetic change to the equation for S. However, by creating a factor with zero mean, we will be able to more sensibly assign it a market price of risk later.

Next, we consider a derivative on the stock $c(S, t)$ which is modeled as a function of S and t. By Ito's lemma we have

$$dc = \mathcal{L}cdt + \sigma S^- c_S dz + \left((c(YS^-) - c(S^-))d\pi - \alpha \mathbb{E}[(c(YS^-) - c(S^-))]dt\right)$$

where

$$\mathcal{L}c = c_t + \mu S^- c_S + \frac{1}{2}\sigma^2(S^-)^2 c_{SS} + \alpha \mathbb{E}[(c(YS^-) - c(S^-))].$$

Now we come to the point where we need to classify all the variables consistently. This is where we run into some difficulty. The problem is how to identify factors. The randomness associated with the jump term does not enter S and c in the same manner. We would like to be able to write a *linear* factor model in the factors dz and $(Y - 1)d\pi$ (or a compensated version of of $(Y - 1)d\pi$). However, in this case that is not possible unless we consider $(c(YS^-) - c(S^-))d\pi$ to be a new factor! However, since this risk is driven by the same Poisson process and jump component Y, we would rather not do this. Yet, at this point we have no choice if we want to force things into the linear factor model framework that we have developed.

What has happened here is that Ito's lemma has *not* produced a *linear* factor model for the derivative in the factors that are driving the stock S. Instead the factors and the randomness are entering in a *nonlinear* fashion.

In any case, let's proceed by considering

$$\left((c(YS^-) - c(S^-))d\pi - \alpha \mathbb{E}[(c(YS^-) - c(S^-))]dt\right) \text{ and } (Y-1)d\pi(\alpha) - \alpha \mathbb{E}[Y-1]dt$$

as two *different* factors. For notational convenience, let's call them

$$
\begin{aligned}
d\psi_1 &= (Y - 1)d\pi(\alpha) - \alpha \mathbb{E}[Y - 1]dt, \\
d\psi_2 &= \left((c(YS^-) - c(S^-))d\pi - \alpha \mathbb{E}[(c(YS^-) - c(S^-))]dt\right).
\end{aligned}
$$

Thus, we arrive at our classification of variables as

Tradables	Underlying Variables	Factors
B, S, c	r_0, S	$dt, dz, d\psi_1, d\psi_2$.

Step 2: Construct a Tradable Table
Our tradable table corresponding to this model is

Prices   Value Change Factor Models

$$
\begin{bmatrix} B \\ S^- \\ -- \\ c \end{bmatrix}
\quad
d\begin{bmatrix} B \\ S \\ -- \\ c \end{bmatrix}
=
\begin{bmatrix} r_0 B \\ (\mu + \alpha \mathbb{E}[Y - 1])S^- \\ -- \\ \mathcal{L}c \end{bmatrix} dt
+
\begin{bmatrix} 0 & 0 & 0 \\ \upsilon S^- & S^- & 0 \\ -- & -- & -- \\ \sigma S^- c_S & 0 & 1 \end{bmatrix}
\begin{bmatrix} dz \\ d\psi_1 \\ d\psi_2 \end{bmatrix}.
$$

Again, we note that it is quite artificial to treat $d\psi_1$ and $d\psi_2$ as separate factors since they are really driven by the same risk, just related to each other in a nonlinear manner. Nevertheless, we will proceed and see what can be done.

Step 3: Apply the Price APT

The Price APT equations take the form,

$$
\begin{bmatrix} r_0 B \\ (\mu + \alpha \mathbb{E}[Y-1]) S^- \\ -- \\ \mathcal{L}c \end{bmatrix} = \begin{bmatrix} B \\ S^- \\ -- \\ c \end{bmatrix} \lambda_0 + \begin{bmatrix} 0 & 0 & 0 \\ \sigma S^- & S^- & 0 \\ -- & -- & -- \\ \sigma S^- c_S & 0 & 1 \end{bmatrix} \begin{bmatrix} \lambda_1 \\ \lambda_2 \\ \lambda_3 \end{bmatrix}.
$$

Now, we could write out the absence of arbitrage equation for the derivative that results from solving the Price APT. This is fairly messy and leaves a degree of freedom since we only have two marketed tradables (B and S) and four market prices of risk. We are in the *underdetermined* case as explained in Chapter 5, Section 5.6.1. However, we note that λ_2 and λ_3 corresponding to $d\psi_1$ and $d\psi_2$ should not be independent of each other since they are really describing the same risk. In any case, we have ourselves a messy situation.

How does Merton untangle himself from this situation? He makes the assumption that all jump risk is diversifiable! That is, the market price of risk of any risk associated with the jump term is zero! This is the same as setting $\lambda_2 = \lambda_3 = 0$. This is a strong assumption!

Note that setting these market prices of risk to zero is reasonable because the factors to which they correspond ($d\psi_1$ and $d\psi_2$) have zero expectation. If we had not compensated the factors to give them zero mean then it would not have made sense to set their market prices of risk to zero. With this assumption, let's follow Merton and see where this leads us.

With $\lambda_2 = \lambda_3 = 0$, the Price APT equations reduce dramatically to

$$
\begin{bmatrix} r_0 B \\ (\mu + \alpha \mathbb{E}[Y-1]) S^- \\ -- \\ \mathcal{L}c \end{bmatrix} = \begin{bmatrix} B \\ S^- \\ -- \\ c \end{bmatrix} \lambda_0 + \begin{bmatrix} 0 \\ \sigma S^- \\ -- \\ \sigma S^- c_S \end{bmatrix} \begin{bmatrix} \lambda_1 \end{bmatrix}.
$$

and we can solve for the remaining market prices of risk as

$$
\lambda_0 = r_0, \qquad \lambda_1 = \frac{\mu + \alpha \mathbb{E}[Y-1]}{\sigma}.
$$

Substituting these values into the final equation for the derivative yields

$$
\mathcal{L}c = r_0 c + \left(\frac{\mu + \alpha \mathbb{E}[Y-1] - r_0}{\sigma} \right) \sigma S^- c_S
$$

which can be reduced to

$$
c_t + (r_0 - \alpha \mathbb{E}[Y-1]) S^- c_S + \frac{1}{2} \sigma^2 (S^-)^2 c_{SS} = r_0 c - \alpha \mathbb{E}[(c(Y S^-) - c(S^-))]. \quad (6.41)
$$

Thus, by making the assumption that jump risk is diversifiable we are able to essentially zero out the nonlinearities that were causing problems and arrive at a nice partial differential/integral absence of arbitrage equation to price the derivative.

6.6.1 Bankruptcy!

Let's consider a special simplifying case in which a stock can go bankrupt and jump to zero. That is, $Y = 0$. The absence of arbitrage equation (6.41) simplifies to

$$c_t + (r_0 + \alpha)Sc_S + \frac{1}{2}\sigma^2 S^2 c_{SS} = (r_0 + \alpha)c. \qquad (6.42)$$

This looks like the standard Black–Scholes equation but the interest rate has been increased by the intensity of the bankruptcy probability! We will see that this same relationship will also appear in defaultable bonds. In fact, this model is really the prototype for defaultable bonds. Of course, we can provide a closed form solution based on the Black–Scholes formula.

Note the following counterintuitive observation. In the Black–Scholes formula, the value of a European call option increases with the risk-free rate. This means that according to our model above, if the rate of bankruptcy increases, then the value of a call option will actually increase! Why is this the case? One of the simplest ways to understand this is to note that a call option allows you to *wait* before purchasing the stock. If the stock has the possibility of going bankrupt, then the value of being able to wait should increase. Thus, an increase in the value of a call option occurs.

Example 6.7 (Bankruptcy!)
Consider a call option with strike price $K = \$10$ and time to expiration $T = 0.3$ on a stock with current price $S = \$10$ and volatility $\sigma = 0.3$. Additionally, assume that this stock has the possibility of jumping into bankruptcy and becoming worthless. Let α be the intensity of the possible jump into bankruptcy. Finally, assume the risk-free rate is $r_0 = 0.05$. According to our analysis above, the value of a call option on this stock is just given by the Black–Scholes formula with the risk-free rate increased to $r_0 + \alpha$.

Figure 6.3 provides a plot of the value of the option as the intensity of bankruptcy α increases from 0 to 0.2. Somewhat counterintuitively, the value of the call option increases quite significantly with the increased likelihood that the stock jumps into bankruptcy.

6.6.2 Lognormal Jumps

When the jump size Y is lognormal, then conditional on the number of jumps that have occurred before expiration, the stock distribution at expiration is lognormal. This situation was explored in Chapter 3, Section 3.3. In this case there is a closed form solution for European call and put options.

To make the jumps lognormal, let $Y = e^Z$ where Z is normally distributed with mean ν_J and standard deviation σ_J. Here, I am using the subscript J to indicate that ν_J and σ_J correspond to the "jump" term. In this case, we define k as the expected value of the jump factor as

$$k = \mathbb{E}[Y - 1] = \mathbb{E}[Y] - 1 = \mathbb{E}[e^Z] - 1 = e^{\mu_J + (0.5)\sigma_J^2} - 1.$$

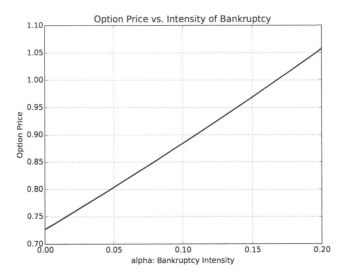

Figure 6.3 Call option value as intensity of bankruptcy is increased.

Finally, we define $\gamma = \ln(\mathbb{E}[Y]) = \ln(1+k)$ and $\alpha' = \alpha(1+k)$. Then, for a European call option with strike K and expiration T, the price under the jump diffusion model is given by the formula

$$c_{JD}(S,t) = \sum_{n=0}^{\infty} \left[\frac{e^{-\alpha'(T-t)}(\alpha'(T-t))^n}{n!} \right] c_{BS}\left(S,t,K,T,r_0 - \alpha k + \frac{n\gamma}{T-t}, \sigma^2 + \frac{n\sigma_J^2}{T-t}\right)$$

where $c_{BS}(S,t,K,T,r_0,\sigma)$ is the Black–Scholes formula for a European call option on a non-dividend paying stock with price S at time t, strike K, expiration T, risk-free rate r_0, and volatility σ. The derivation of this can be found in Merton's work [40].

Note that the jump diffusion formula appears to be a combination of the Black–Scholes formula and the solution under Poisson dynamics. The key is that under this jump diffusion model, conditional on the number of jumps, the price distribution at expiration is lognormal, indicating that the solution should "look like the Black–Scholes formula." Again, see Section 3.3 of Chapter 3 where the dynamics for a jump diffusion process were introduced. The only question is how many jumps have occurred. Therefore, the solution is basically that conditioned on the number of jumps that have occurred, the call price should be given by the Black–Scholes formula. That is, for each possible number of jumps, we have a Black–Scholes formula, but weighted by the probability of that number of jumps occurring, which is Poisson. Simple!

Example 6.8 (Jump Diffusion Volatility "Smile")
Consider a non-dividend paying stock that follows jump diffusion dynamics. Assume the current price of the stock is $S = \$10.0$, the volatility associated with the

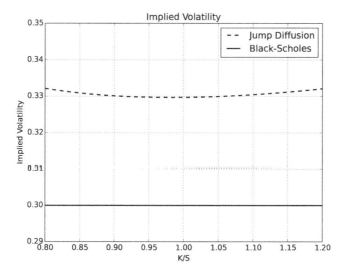

Figure 6.4 Implied volatility "smile" under jump diffusion dynamics.

Brownian factor is $\sigma = 0.3$, and the jump term has intensity $\alpha = 2$ with jump parameters $\nu_J = 0$ and $\sigma_J = 0.1$. Roughly speaking, when no jumps are occurring, the stock moves like a geometric Brownian motion with 30% volatility. Because the jump intensity is $\alpha = 2$, we expect to see two lognormal jumps a year. The log-mean of each jump is $\nu_J = 0$ and the log-standard deviation is 0.1, meaning the jump size is about 10% on average.

Consider an option with $T = 0.3$ to expiration (that is, the current time is $t = 0$ and the expiration time is $T = 0.3$). Using Merton's jump diffusion formula, we can compute the price of this option for various values of the strike price K.

One way to understand how this model differs from the Black–Scholes formula (6.7) is to plot the implied volatility curve. Recall that implied volatility was introduced in Section 6.1.2. That is, for a given strike price K, we compute the value of the option under the jump diffusion model. We then ask what volatility value σ would have to be used in the Black–Scholes formula of (6.7) to obtain this same price for the option. This value is the implied volatility, σ_i.

Figure 6.4 provides a plot of the implied volatility for the jump diffusion model under the above parameter values as a function of the ratio of the strike to the current price of the stock K/S. For reference, the implied volatility of the options when no jumps are present (i.e., under Black–Scholes assumptions) is also provided as the solid line. The plot of the implied volatility under the jump diffusion model shows a slight "smile." That is, as the strike price K deviates from the current price of the stock S, causing the option to become more in- or out-of-the-money, the implied volatility increases. This indicates that far in- or out-of-the-money options are more affected (and priced higher) than at the money options when jumps are present.

6.7 EXCHANGE ONE ASSET FOR ANOTHER

In this section we derive the equation for an option that allows the holder to exchange one asset for another at some expiration date. This option was analyzed by Margrabe [37], and is one of my favorite derivatives. Why? Because many other derivatives can be thought of as exchanging two different assets. For instance, can a standard European call option be thought of as exchanging one asset for another? What are the two assets being exchanged? (See Exercise 6.15.)

Step 1: Model and Classify Variables

We start with a risk-free bond B, and the prices of two other assets S_1 and S_2 that are assumed to be non-dividend paying. They evolve according to the stochastic differential equations,

$$dB = r_0 B dt, \tag{6.43}$$
$$dS_1 = \mu_1 S_1 dt + \sigma_1 S_1 dz_1, \tag{6.44}$$
$$dS_2 = \mu_2 S_2 dt + \sigma_2 S_2 dz_2, \tag{6.45}$$

where we also assume that dz_1 and dz_2 are correlated with $\mathbb{E}[dz_1 dz_2] = \rho dt$.

In this case, we are considering a derivative that gives the holder the option to exchange S_1 for S_2 at a fixed expiration date T. Thus, we assume the derivative c is a function of S_1, S_2, and t, as $c(S_1, S_2, t)$. By Ito's lemma, the derivative follows the stochastic differential equation,

$$dc = \mathcal{L}_1 c dt + +\sigma_1 S_1 c_{S_1} dz_1 + \sigma_2 S_2 c_{S_2} dz_2 \tag{6.46}$$

where

$$\mathcal{L}_1 c = (c_t + \mu_1 S_1 c_{S_1} + \mu_2 S_2 c_{S_2} + \frac{1}{2}\sigma_1^2 S_1^2 c_{S_1 S_1} + \frac{1}{2}\sigma_2^2 S_2^2 c_{S_2 S_2} + \rho \sigma_1 \sigma_2 S_1 S_2 c_{S_1 S_2}).$$

Our classification of variables is

$$
\begin{array}{ccc}
\text{Tradables} & \text{Underlying Variables} & \text{Factors} \\
B, S_1, S_2, c & r_0, S_1, S_2 & dt, dz_1, dz_2.
\end{array}
$$

Step 2: Construct a Tradable Table

The tradable table corresponding to this setup is

$$
\begin{array}{cc}
\text{Prices} & \text{Value Change Factor Models} \\
\end{array}
$$

$$
\begin{bmatrix} B \\ S_1 \\ S_2 \\ -- \\ c \end{bmatrix}
\quad
d \begin{bmatrix} B \\ S_1 \\ S_2 \\ -- \\ c \end{bmatrix}
=
\begin{bmatrix} r_0 B \\ \mu_1 S_1 \\ \mu_2 S_2 \\ -- \\ \mathcal{L}c \end{bmatrix} dt
+
\begin{bmatrix} 0 & 0 \\ \sigma_1 S_1 & 0 \\ 0 & \sigma_2 S_2 \\ -- & -- \\ \sigma_1 S_1 c_{S_1} & \sigma_2 S_2 c_{S_2} \end{bmatrix}
\begin{bmatrix} dz_1 \\ dz_2 \end{bmatrix}.
$$

Step 3: Apply the Price APT

Finally, applying the Price APT results in

$$
\begin{bmatrix} r_0 B \\ \mu_1 S_1 \\ \mu_2 S_2 \\ -- \\ \mathcal{L}c \end{bmatrix} = \begin{bmatrix} B \\ S_1 \\ S_2 \\ -- \\ c \end{bmatrix} \lambda_0 + \begin{bmatrix} 0 & 0 \\ \sigma_1 S_1 & 0 \\ 0 & \sigma_2 S_2 \\ -- & -- \\ \sigma_1 S_1 c_{S_1} & \sigma_2 S_2 c_{S_2} \end{bmatrix} \begin{bmatrix} \lambda_1 \\ \lambda_2 \end{bmatrix}.
$$

Using the first three equations to solve for λ_0, λ_1 and λ_2 yields

$$
\lambda_0 = r_0, \qquad \lambda_1 = \frac{\mu_1 - r_0}{\sigma_1}, \qquad \lambda_2 = \frac{\mu_2 - r_0}{\sigma_2}.
$$

Substituting these into the final equation leads to

$$
\mathcal{L}c = r_0 c + \frac{\mu_1 - r_0}{\sigma_1} \sigma_1 S_1 c_{S_1} + \frac{\mu_2 - r_0}{\sigma_2} \sigma_2 S_2 c_{S_2}
$$

which upon rearrangement becomes

$$
c_t + r_0 S_1 c_{S_1} + r_0 S_2 c_{S_2} + \frac{1}{2} \sigma_1^2 S_1^2 c_{S_1 S_1} + \frac{1}{2} \sigma_2^2 S_2^2 c_{S_2 S_2} + \rho \sigma_1 \sigma_2 S_1 S_2 c_{S_1 S_2} = r_0 c. \quad (6.47)
$$

This is the absence of arbitrage equation satisfied by a derivative dependent on S_1 and S_2.

6.7.1 Closed Form Solution

There is a surprisingly simple closed form solution for a European option to exchange one asset for another. Assume that the derivative is the option to exchange asset S_1 for asset S_2 at time T. You will use this option if S_2 is greater in value than S_1, in which case exchanging will net you the difference, $S_2 - S_1$. On the other hand, if S_1 is greater in value than S_2 at expiration, then the option is worthless. Thus, the payoff function of this option is

$$
c(S_1, S_2, T) = \max \{ S_2 - S_1, 0 \}. \quad (6.48)
$$

The closed form solution for this option turns out to be

$$
c(S_1, S_2, t) = S_2 \Phi(\hat{d}_1) - S_1 \Phi(\hat{d}_2) \quad (6.49)
$$

where

$$
\hat{d}_1 = \frac{\ln(S_2/S_1) + \frac{1}{2}\hat{\sigma}^2 (T - t)}{\hat{\sigma}\sqrt{T - t}} \quad (6.50)
$$

$$
\hat{d}_2 = \hat{d}_1 - \hat{\sigma}\sqrt{T - t} \quad (6.51)
$$

$$
\hat{\sigma} = \sqrt{\sigma_1^2 + \sigma_2^2 - 2\rho\sigma_1\sigma_2}. \quad (6.52)
$$

Note that this solution looks very much like the Black–Scholes formula (6.7). Here is some intuition into why. Let's take the payoff function and rewrite it slightly as

$$c(S_1, S_2, T) = \max\{S_2 - S_1, 0\} = S_1 \max\left\{\frac{S_2}{S_1} - 1, 0\right\}.$$

Written this way, the term $\max\{\frac{S_2}{S_1} - 1, 0\}$ looks like the payoff of a call option with strike price 1 on the asset $\frac{S_2}{S_1}$. In fact, dividing S_2 by S_1 is essentially changing units in order to value S_2 in units of S_1. Said differently, the ratio $\frac{S_2}{S_1}$ is just the number of shares of S_1 that it takes to purchase S_2. Therefore, in units of S_1, this is just a call option on S_2 with strike 1. The multiplication on the left by S_1 just converts the call option back to units of dollars. Therefore, we shouldn't be surprised that the solution looks like the Black–Scholes formula, because this is really just a call option in a different set of units.

When we change units by denominating quantities in terms of another asset (such as S_1 in this case), we refer to this as a *change of numeraire*, and we call the asset S_1 the *numeraire* asset. The numeraire is the "currency" that we use to express the value of all other assets. Needless to say, some problems (such as this one) are easier to solve in the "right" set of units.

6.7.2   Using a Different Currency (Change of Numeraire)

As mentioned above, it is sometimes more convenient to denominate asset prices in a different set of units or currency. As an exercise, let's do this in exchange one asset for another. Let S_1 be the numeraire. That is, we will denominate all quanitites in units of the S_1 currency. So now if someone asks, "What is you car worth?" you respond, "500 shares of S_1!"

Step 1: Model and Classify Variables
In our new currency, the risk-free bond is expressed in units of S_1 as $\frac{B}{S_1}$, and by Ito's lemma it satisfies

$$d\frac{B}{S_1} = r_0\frac{B}{S_1}dt - \frac{B}{S_1^2}(\mu_1 S_1 dt + \sigma_1 S_1 dz_1) + \frac{B}{S_1}\sigma_1^2 dt.$$

To indicate that we are denominating prices using the "S_1" currency, let's use a superscript "1." For example, we will write $B^1 = \frac{B}{S_1}$. From above, we can write its dynamics as

$$dB^1 = \left(r_0 - \mu_1 + \sigma_1^2\right)B^1 dt - \sigma_1 B^1 dz_1.$$

We see that the "risk-free asset" is no longer risk free when denominated in units of S_1. On the other hand, when we denominate S_1 in units of S_1 we just have the constant 1. Again, using the superscript notation, $S_1^1 = \frac{S_1}{S_1} = 1$, which leads to

$$dS_1^1 = 0.$$

Thus, S_1^1 becomes the new "risk-free" asset!

Next, we denominate S_2 in terms of S_1 as $S_2^1 = \frac{S_2}{S_1}$ and apply Ito's lemma to obtain

$$d\frac{S_2}{S_1} = \frac{S_2}{S_1}\left(\mu_2 dt + \sigma_2 dz_2\right) - \frac{S_2}{S_1^2}\left(\mu_1 S_1 dt + \sigma_1 S_1 dz_1\right) - \frac{1}{S_1^2}\rho\sigma_1\sigma_2 S_1 S_2 dt + \frac{S_2}{S_1^3}\sigma_1^2 S_1^2 dt$$

or, using the superscript notation,

$$dS_2^1 = \left(\mu_2 - \mu_1 - \rho\sigma_1\sigma_2 + \sigma_1^2\right)S_2^1 dt + \sigma_2 S_2^1 dz_2 - \sigma_1 S_2^1 dz_1.$$

Finally, exchanging one asset for another can be considered a derivative that depends on S_2^1 and is thus written as $c^1(S_2^1, t)$. Again, we are using the superscript notation for the derivative c^1 to indicate that we are also pricing the derivative in units of S_1. That is, $c^1 = \frac{c}{S_1}$. By Ito's lemma we have

$$dc^1 = \left[c_t^1 + \left(\mu_2 - \mu_1 - \rho\sigma_1\sigma_2 + \sigma_1^2\right)S_2^1 c_{S_2^1}^1 \right.$$
$$+ \frac{1}{2}\left(\sigma_2^2 + \sigma_1^2 - 2\rho\sigma_1\sigma_2\right)(S_2^1)^2 c_{S_2^1 S_2^1}^1 \Bigg] dt$$
$$+ \sigma_2 S_2^1 c_{S_2^1}^1 dz_2 - \sigma_1 S_2^1 c_{S_2^1}^1 dz_1.$$

To write this more succinctly, we define

$$\hat{\mu} = \mu_2 - \mu_1 - \rho\sigma_1\sigma_2 + \sigma_1^2,$$
$$\hat{\sigma}^2 = \sigma_2^2 + \sigma_1^2 - 2\rho\sigma_1\sigma_2, \tag{6.53}$$
$$\mathcal{L}c^1 = c_t^1 + \hat{\mu}S_2^1 c_{S_2^1}^1 + \frac{1}{2}\hat{\sigma}^2(S_2^1)^2 c_{S_2^1 S_2^1}^1,$$

and note that

$$dc^1 = \mathcal{L}c^1 dt - \sigma_1 S_2^1 c_{S_2^1}^1 dz_1 + \sigma_2 S_2^1 c_{S_2^1}^1 dz_2.$$

We can classify the variables under our new S_1 currency as

Tradables	Underlying Variables	Factors
$B^1, S_1^1 = 1, S_2^1, c^1$	r_0, S_2^1	$dt, dz_1, dz_2.$

In fact, as we proceed we will see that B^1 and r_0 don't really matter with S^1 as numeraire.

Step 2: Construct a Tradable Table
The tradable table corresponding to this set up is

Prices   Value Change Factor Models

$$\begin{bmatrix} B^1 \\ S_1^1 \\ S_2^1 \\ -- \\ c^1 \end{bmatrix} \quad d\begin{bmatrix} B^1 \\ S_1^1 \\ S_2^1 \\ -- \\ c^1 \end{bmatrix} = \begin{bmatrix} \left(r_0 - \mu_1 + \sigma_1^2\right)B^1 \\ 0 \\ \hat{\mu}S_2^1 \\ -- \\ \mathcal{L}c^1 \end{bmatrix} dt + \begin{bmatrix} -\sigma_1 B^1 & 0 \\ 0 & 0 \\ -\sigma_1 S_2^1 & \sigma_2 S_2^1 \\ -- & -- \\ -\sigma_1 S_2^1 c_{S_2^1}^1 & \sigma_2 S_2^1 c_{S_2^1}^1 \end{bmatrix}\begin{bmatrix} dz_1 \\ dz_2 \end{bmatrix}.$$

Step 3: Apply the Price APT

Finally, applying the Price APT gives

$$
\begin{bmatrix}
(r_0 - \mu_1 + \sigma_1^2)\,B^1 \\
0 \\
\hat{\mu}S_2^1 \\
-- \\
\mathcal{L}c^1
\end{bmatrix}
=
\begin{bmatrix}
B^1 \\
S_1^1 \\
S_2^1 \\
-- \\
c^1
\end{bmatrix}
\lambda_0
+
\begin{bmatrix}
-\sigma_1 B^1 & 0 \\
0 & 0 \\
-\sigma_1 S_2^1 & \sigma_2 S_2^1 \\
-- & -- \\
-\sigma_1 S_2^1 c_{S_2^1}^1 & \sigma_2 S_2^1 c_{S_2^1}^1
\end{bmatrix}
\begin{bmatrix}
\lambda_1 \\
\lambda_2
\end{bmatrix}.
$$

From the second equation we have $\lambda_0 = 0$. The first equation then allows us to solve for λ_1 as

$$
\lambda_1 = -\frac{(r_0 - \mu_1 + \sigma_1^2)}{\sigma_1}.
$$

From the third equation we have

$$
\hat{\mu} = -\sigma_1 \lambda_1 + \sigma_2 \lambda_2. \tag{6.54}
$$

Moving to the fourth equation for the derivative, we obtain

$$
\mathcal{L}c^1 = [-\sigma_1 \lambda_1 + \sigma_2 \lambda_2]\, S_2^1 c_{S_2^1}^1.
$$

Substituting for $\mathcal{L}c^1$ on the left and using equation (6.54) on the right gives

$$
c_t^1 + \hat{\mu} S_2^1 c_{S_2^1}^1 + \frac{1}{2}\hat{\sigma}^2 (S_2^1)^2 c_{S_2^1 S_2^1}^1 = \hat{\mu} S_2^1 c_{S_2^1}^1
$$

which reduces to the final equation

$$
c_t^1 + \frac{1}{2}\hat{\sigma}^2 (S_2^1)^2 c_{S_2^1 S_2^1}^1 = 0. \tag{6.55}
$$

If we compare this equation with the standard Black–Scholes equation (6.6), we see that it matches by setting $r_0 = 0$ and $\sigma = \hat{\sigma}$. Therefore, it is no surprise that we have a closed form solution!

6.7.3   Recovering the Closed Form Solution

As detailed previously, the payoff of exchange one asset for another when denominated in units of S_1 is

$$
c^1(S_2^1, T) = \max\{S_2^1 - 1, 0\}.
$$

Combining this with the absence of arbitrage equation (6.55), we see that the solution is just the Black–Scholes formula for a standard European call option with $r_0 = 0$, $\sigma = \hat{\sigma}$, $S = S_2^1$, and $K = 1$, given by

$$
c^1(S_2^1, t) = S_2^1 \Phi(\hat{d}_1) - \Phi(\hat{d}_2) \tag{6.56}
$$

with

$$\hat{d}_1 = \frac{\ln(S_2^1) + (\frac{1}{2}\hat{\sigma}^2)(T-t)}{\hat{\sigma}\sqrt{T-t}},$$
$$\hat{d}_2 = \hat{d}_1 - \hat{\sigma}\sqrt{T-t}.$$

Now, this solution is in units of S_1. The final step is to convert this to the original units of dollars. To do this we simply multiply the solution by S_1. That is, by definition, $c^1 = \frac{c}{S_1}$, so $c = S_1 c^1$. This converts from shares of S_1 back into the original dollar currency. So, the answer in units of dollars is

$$
\begin{aligned}
c(S_1, S_2, t) &= S_1 c^1(S_2^1, t) \\
&= S_1 \left(S_2^1 \Phi(\hat{d}_1) - \Phi(\hat{d}_2) \right) \\
&= S_2 \Phi(\hat{d}_1) - S_1 \Phi(\hat{d}_2)
\end{aligned}
\tag{6.57}
$$

and we are done!

We will see this change of unit trick later in the context of interest rate derivatives when we consider the LIBOR Market Model in Chapter 7. The idea of changing to a different "currency" is very important, so take the time to understand what we did. In particular, when you see quantities divided by the price of a tradable, it is often a good idea to change to units of that tradable and interpret it as your new currency.

Example 6.9 (Pricing an Option to Exchange)
Consider the option to exchange stock S_1 for stock S_2 at time $T = 0.25$. Assume that the volatilities of the stocks are $\sigma_1 = 0.3$ and $\sigma_2 = 0.4$, respectively, while the correlation coefficient between the stocks is $\rho = 0.6$. The current prices of the stocks are $S_1 = \$10$ and $S_2 = \$12$.

To compute the value of this option, first we calculate $\hat{\sigma}$ from equation (6.53) as

$$
\begin{aligned}
\hat{\sigma} &= \sqrt{\sigma_1^2 + \sigma_2^2 - 2\rho\sigma_1\sigma_2} \\
&= \sqrt{0.3^2 + 0.4^2 - 2(0.6)(0.3)(0.4)} = 0.3256.
\end{aligned}
$$

We then calculate $\hat{d}_1$ and $\hat{d}_2$ as

$$
\begin{aligned}
\hat{d}_1 &= \frac{\ln(12/10) + (0.5)(0.3256^2)(0.25)}{(0.3256)(\sqrt{0.25})} = 1.2014, \\
\hat{d}_2 &= 1.2014 - (0.3256)(\sqrt{0.25}) = 1.0386.
\end{aligned}
$$

Using these values in the formula for the option to exchange (6.57) leads to

$$c = (12)\Phi(1.2014) - (10)\Phi(1.0386) = 2.1174.$$

6.8 STOCHASTIC VOLATILITY

Stochastic volatility models dispense with the assumption that the volatility σ is constant, and instead allow it to follow its own stochastic differential equation. Early influential work in this area includes [29] and [26], and there now exist entire books devoted to this topic alone, such as [1] and [20]. Stochastic volatility models play an important role in option pricing because they more accurately model real asset price movements, and they capture important features such as volatility smiles (see Example 6.6.2 and [28, 53]). We will see that stochastic volatility models are incomplete in that volatility is not tradable, but is a source of risk affecting the value of an option. This means that we won't arrive at a unique price and the solution will be left in terms of a market price of volatility risk.

Step 1: Model and Classify Variables

First, we can list the relevant equations for underlying variables. In this case, we model a risk-free bond B, a stock S, and an instantaneous variance variable v whose square root determines the volatility of the stock,

$$dB = r_0 B dt, \tag{6.58}$$
$$dS = \mu S dt + \sqrt{v} S dz_1, \tag{6.59}$$
$$dv = a(v) dt + b(v) dz_2, \tag{6.60}$$

where we assume that z_1 and z_2 are correlated with $\mathbb{E}[dz_1 dz_2] = \rho dt$. In this case, $a(v)$ and $b(v)$ will take some specific form to capture the dynamics of v. For example, if v is mean reverting we would choose $a(v) = -k(v - \theta)$. In fact, we may want to model v as a CIR process (see Chapter 3, Section 3.5) or some other sensible process. At this point, I am not concerned with the specific form of the dynamics, and for notational convenience, I won't even show the dependence of a and b on v and other variables, but note that it can be there.

We would like to price a derivative on the stock. Note that our derivative can depend on S, t, and v, as in $c(S, v, t)$. By Ito's lemma we have

$$dc = \mathcal{L} c dt + \sqrt{v} S c_S dz_1 + b c_v dz_2 \tag{6.61}$$

where

$$\mathcal{L} c = (c_t + \mu S c_S + a c_v + \frac{1}{2} v S^2 c_{SS} + \frac{1}{2} b^2 c_{vv} + \rho b \sqrt{v} S c_{Sv}). \tag{6.62}$$

Our classification of variables is

Tradables	Underlying Variables	Factors
B, S, c	r_0, S, v	$dt, dz_1, dz_2.$

Step 2: Construct a Tradable Table

Collecting the dynamics of the tradables from above leads to the following tradable table

$$
\text{Prices} \qquad \text{Value Change Factor Models}
$$

$$
d\begin{bmatrix} B \\ S \\ -- \\ \iota \end{bmatrix} \begin{bmatrix} B \\ S \\ -- \\ \iota \end{bmatrix} = \begin{bmatrix} r_0 B \\ \mu S \\ -- \\ \mathcal{L}\iota \end{bmatrix} dt + \begin{bmatrix} 0 & 0 \\ \sqrt{v}S & 0 \\ -- & -- \\ \sqrt{v}S\iota_S & b\iota_v \end{bmatrix} \begin{bmatrix} dz_1 \\ dz_2 \end{bmatrix}.
$$

Note that the instantaneous variance variable v is not tradable.

Step 3: Apply the Price APT

Applying the Price APT equation results in

$$
\begin{bmatrix} r_0 B \\ \mu S \\ -- \\ \mathcal{L}c \end{bmatrix} = \begin{bmatrix} B \\ S \\ -- \\ c \end{bmatrix} \lambda_0 + \begin{bmatrix} 0 & 0 \\ \sqrt{v}S & 0 \\ -- & -- \\ \sqrt{v}Sc_S & bc_v \end{bmatrix} \begin{bmatrix} \lambda_1 \\ \lambda_2 \end{bmatrix}, \qquad (6.63)
$$

and solving for the market prices of risks λ_0 and λ_1 gives

$$
\lambda_0 = r_0, \quad \lambda_1 = \frac{\mu - r_0}{\sqrt{v}}. \qquad (6.64)
$$

Note that λ_2 is left as an unknown. Substituting these into the final equation results in

$$
c_t + r_0 S c_S + (a - \lambda_2 b)c_v + \frac{1}{2}vS^2 c_{SS} + \frac{1}{2}b^2 c_{vv} + \rho b\sqrt{v}S c_{Sv} = r_0 c. \qquad (6.65)
$$

The incompleteness of the market shows up as the unknown market price of risk λ_2, which corresponds to the untraded volatility risk factor dz_2. Thus, in this case there is not a unique absence of arbitrage price. The price will be a function of whatever value is chosen for the market price of risk λ_2.

Under specific choices of a and b for a European call option, equation (6.65) has a fairly convenient solution via transform methods. I won't pursue this in detail here. The interested reader is encouraged to consult some of the texts in the references such as [28] or [53].

6.9   SUMMARY

This chapter showed the power of applying the factor model approach to arbitrage pricing in the context of equity derivative models. We systematically attacked each situation using our three step procedure. The hope is that you have gained enough familiarity with the approach that you can now apply it to any new problem you encounter. But the journey doesn't stop with equity derivatives. In the next chapter we will explore the factor model approach to interest rate and credit derivatives.

EXERCISES

6.1 Consider a non-dividend paying stock with current price $S(0) = \$25$ and volatility $\sigma = 30\%$. The risk-free rate is $r_0 = 4\%$. According to the Black–Scholes formula, what are the prices of a European call and put option with strike price $K = \$26$ and expiration $T = 0.2$?

6.2 Consider a non-dividend paying stock with current price $S(0) = \$12$ and volatility $\sigma = 20\%$. The risk-free rate is $r_0 = 2\%$. According to the Black–Scholes formula, what are the prices of a European call and put option with strike price $K = \$11$ and expiration $T = 0.3$?

6.3 Consider a stock with current price $S(0) = \$10$ and volatility $\sigma = 40\%$. Assume the stock pays a continuous dividend at a rate of $q = 3\%$. The risk-free rate is $r_0 = 3\%$. According to the Black–Scholes–Merton formula, what are the prices of European call and European put options with strike price $K = \$9.50$ and expiration $T = 0.30$?

6.4 Consider a stock with current price $S(0) = \$55$ and volatility $\sigma = 25\%$. Assume the stock pays a continuous dividend at a rate of $q = 1\%$. The risk-free rate is $r_0 = 5\%$. According to the Black–Scholes–Merton formula, what are the prices of European call and European put options with strike price $K = \$57$ and expiration $T = 0.15$?

6.5 Consider a stock with current price $S(0) = \$50$ and volatility $\sigma = 25\%$. The risk-free rate is $r_0 = 6\%$. Assume the stock will pay a cash dividend at time $\tau = 0.15$ in an amount of $D_\tau = \$1.20$. What are the prices of European call and European put options with strike price $K = \$49$ and expiration $T = 0.40$?

6.6 Consider a stock with current price $S(0) = \$25$ and volatility $\sigma = 35\%$. The risk-free rate is $r_0 = 3\%$. Assume the stock will pay a cash dividend at time $\tau = 0.05$ in an amount of $D_\tau = \$0.65$. What are the prices of European call and European put options with strike price $K = \$25.5$ and expiration $T = 0.20$?

6.7 Consider a futures contract with expiration $T = 0.5$. The current futures price is $f(0) = \$35$ and the volatility of the futures price is $\sigma = 30\%$. The risk-free rate is $r_0 = 4\%$. What are the prices of European call and European put options on the futures contract with strike price $K = \$37$ and expiration $T = 0.5$?

6.8 Consider a futures contract with expiration $T = 0.25$. The current futures price is $f(0) = \$12$ and the volatility of the futures price is $\sigma = 40\%$. The risk-free rate is $r_0 = 2\%$. What are the prices of European call and European put options on the futures contract with strike price $K = \$11$ and expiration $T = 0.25$?

6.9 Consider a European call option on a non-dividend paying stock with current price $S(0) = \$10$, strike $K = \$10$, and expiration $T = 0.2$. The risk-free rate is 4%. Compute the price of the call for the volatilities of $\sigma = 20\%$, 30%, and 40%. Indicate whether the value of the call options increases or decreases with increasing volatility.

6.10 Repeat Exercise 6.9 for a European put option instead of a European call option. Indicate whether the value of the put option increases or decreases with increasing volatility.

6.11 Compute the implied volatility of a call option with market price $1.36, strike price $K = 20.50$, and expiration $T = 0.3$ on a non-dividend paying stock. The current price of the stock is $S(0) = \$20$. The risk-free rate is $r_0 = 2\%$.

6.12 Compute the implied volatility of a put option with market price 0.45, strike price $K = 12.5$, and expiration $T = 0.25$ on a non-dividend paying stock. The current price of the stock is $S(0) = \$12$. The risk-free rate is $r_0 = 5\%$.

6.13 Stock S_1 has current price $S_1(0) = \$10$ and volatility $\sigma_1 = 0.2$. Stock S_2 has current price $S_2(0) = \$9.80$ and volatility $\sigma_2 = 0.3$. The correlation coefficient between S_1 and S_2 is $\rho = 0.3$. What is the price of the option to exchange S_1 for S_2 with expiration $T = 0.2$?

6.14 Stock S_1 has current price $S_1(0) = \$28$ and volatility $\sigma_1 = 0.3$. Stock S_2 has current price $S_2(0) = \$30$ and volatility $\sigma_2 = 0.25$. The correlation coefficient between S_1 and S_2 is $\rho = -0.1$. What is the price of the option to exchange S_1 for S_2 with expiration $T = 0.4$?

6.15 Show that the Black–Scholes formula (6.7) is a special case of exchange one asset for another where S_2 is the stock $S(t)$ and S_1 is a risk-free bond $B(t|T) = Ke^{-r_0(T-t)}$ that pays exactly K at expiration T. In particular, show that the exchange one asset for another formula (6.49) reduces to the Black–Scholes formula (6.7) in this case.

6.16 Exchange One Asset for Another.

In this problem you will derive the formula for an option to exchange one asset for another by reducing the absence of arbitrage equation to a standard Black–Scholes partial differential equation for a call option on geometric Brownian motion.

(a) Let S_1 and S_2 be two assets whose dynamics are given by

$$
\begin{aligned}
dS_1 &= \mu_1 S_1 dt + \sigma_1 S_1 dz_1, & \text{(P6.1)} \\
dS_2 &= \mu_2 S_2 dt + \sigma_2 S_2 dz_2, & \text{(P6.2)}
\end{aligned}
$$

where $\mathbb{E}[dz_1 dz_2] = \rho dt$ and assume that a risk-free asset exists with constant interest rate r_0. Write the absence of arbitrage equation for an option to exchange S_1 for S_2 at expiration T. (This was essentially already done for you in equation (6.47).)

(b) Consider the change of variable $v = S_2/S_1$ and assume that the solution of the above absence of arbitrage partial differential equation is of the form $c(S_1, S_2, t) = S_1 f(v, t)$. Using these substitutions, write the partial differential equation from (a) in terms of S_1, t, f, and v. What equation must $f(v, t)$ satisfy? What is the appropriate boundary condition for a European option to exchange S_1 for S_2 in terms of the new variables?

(c) Using the Black–Scholes formula, write a formula for the value of an option to exchange S_1 for S_2. (Recall that the payoff is $\max\{S_2 - S_1, 0\}$.)

6.17 Consider a European call option on a non-dividend paying stock with stochastic volatility. This time we model the instantaneous variance with a Poisson process. The idea is that most of the time the instantaneous variance is σ_l^2, but at random times the market goes wild and the instantaneous variance jumps by an amount b. After the jump, the instantaneous variance exponentially decays back to its normal level σ_l^2. First write the relevant dynamics for this problem, then derive the absence of arbitrage equation for the price of the option. Again, this absence of arbitrage equation can be in terms of a market price of risk.

(Hint: your dynamics should look like this:

$$
\begin{aligned}
dS &= \mu S dt + \sqrt{v} S dz \\
dv &= a(\sigma_l^2 - v^-) dt + b d\pi(\alpha),
\end{aligned}
$$

where π is a Poisson process with intensity α.)

CHAPTER 7

Interest Rate and Credit Derivatives

I N this chapter, we apply the factor approach to interest rate and credit derivatives. As in the previous chapter, the emphasis is on showing that the absence of arbitrage equations follow from our factor model based three step approach.

One of the central issues in interest rate derivatives is absence of arbitrage modeling of the so-called *term structure of interest rates*. We begin with a review of alternative choices for describing the term structure of interest rates. These provide motivation for the selection of underlying variables when we turn to the task of constructing absence of arbitrage models.

We will see that there are two main approaches, which correspond to either driving the term structure with a random short rate process or with forward rates. Short rate based models are considered first, where we focus on deriving absence of arbitrage equations and formulas for zero-coupon bonds. Then we turn our attention to models based on forward rates as underlying variables. To better understand the use of forward rates as underlying variables, we first take a slight detour into derivatives on forward contracts. The results for forward rates then follow quite directly, including the well-known LIBOR Market Model based caplet formula. Finally, we provide a brief introduction to reduced-form modeling of credit derivatives, where links with Merton's jump diffusion model are highlighted.

7.1 INTEREST RATE MODELING AND DERIVATIVES

Before getting started, let me make a few comments about interest rate modeling and derivatives. For any derivative pricing problem, the first challenge before being able to price a derivative is the calibration phase as in Figure 5.3 of Chapter 5. For many equity derivatives, the calibration phase does not play prominently into the analysis. In fact, in many equity derivative models the market is either complete, in which case the calibration phase of determining market prices of risk is quite trivial, or the market is incomplete, in which case the Price APT equations are underdetermined and we are left with a degree of freedom in choosing market prices of risk.

Interest rate and so-called term structure models tend to be the opposite. That is, we may use a single factor (thus there is one market price of risk λ), but have many marketed tradables. This often makes the APT equations overdetermined! Thus, the calibration phase is in determining a λ value that best fits all the marketed tradables. (In theory if there isn't a perfect fit, then an arbitrage is available. However, in practice, we often recognize that our model is too simple and try to find a best fit λ.) Once calibration is completed, pricing a derivative proceeds by solving the appropriate absence of arbitrage equation with boundary conditions using the calibrated λ.

7.2 MODELING THE TERM STRUCTURE OF INTEREST RATES

As background, I will provide a brief description of the so-called *term structure of interest rates*. Those of you who require further background in this area are referred to the early chapters of the excellent book by Luenberger [36].

One of the central challenges in dealing with interest rates is how to model the term structure of interest rates. The phrase *term structure* indicates that a quantity varies depending on the time span (or term) over which it is defined. In this case, the interest rate on a loan is a function of the maturity of the loan, or how long it will be until the loan is repaid. Typically, if someone wants to borrow money short term the interest rate will be lower than if someone wants to borrow long term, such as for 30 years. The fact that there are different interest rates corresponding to different maturity times of loans is what is known as the *term structure of interest rates*. The challenge is to create a stochastic model of this term structure that accurately captures its movement and doesn't allow one to arbitrage off of assets that depend on these interest rates.

In interest rate modeling, there are four standard and equivalent descriptions of the term structure of interest rates. They will play an important role in our development of absence of arbitrage models for the term structure, so we briefly describe them next.

7.2.1 Spot Rates

The most direct way to represent the term structure of interest rates is with a *spot rate curve*, as shown in Figure 7.1, which simply displays the market rate of interest for each maturity time. That is, $r(t|T)$ is the interest rate that you would pay on a risk-free loan where the money is borrowed at the current time t and repaid completely at time T with no intermediate payments. Note that in this description, the maturity time T determines the rate $r(t|T)$, and thus, as depicted in Figure 7.1, we have an entire curve of rates corresponding to different maturity times T.

The interest rate can be quoted using various conventions. In this introduction, we will use the so-called continuous compounding convention, which means that if

The spot rate curve at time t.

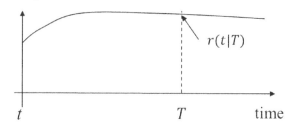

Figure 7.1 Term structure of interest rates as represented by the spot rate curve.

you borrow \$1 at time t and have to repay it at time T, you will have to repay an amount of $e^{r(t|T)(T-t)}$. Likewise, if you agree to pay the bank \$1 at time T, it will allow you to borrow an amount $e^{-r(t|T)(T-t)}$ at time t. There are other conventions that can be used to quote interest rates that will be encountered later in this chapter. But, to introduce you to the concepts related to describing the term structure, I will stick with this continuous compounding convention. The other conventions will be described as needed. Readers interested in gaining broader exposure to various interest rate quoting conventions are referred to the references [18, 36].

7.2.2   Zero-Coupon Bonds

An alternative way to describe the term structure of interest rates is to specify the prices of zero-coupon bonds that pay exactly \$1 at maturity time T. We denote the price of such a zero-coupon bond as $B(t|T)$, where t is the current time and T is a maturity time when the single dollar will be paid. A cash flow diagram corresponding to such a zero-coupon bond is shown in Figure 7.2.

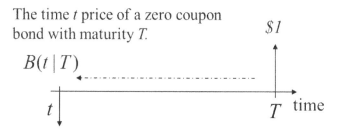

Figure 7.2 Cash flow diagram for zero-coupon bond, $B(t|T)$.

Given the price of a zero-coupon bond $B(t|T)$, we can compute the implied spot rate for its maturity time $r(t|T)$ using the fact that the price $B(t|T)$ should be equal to the payoff of \$1 at time T discounted back to the current time t using the spot rate $r(t|T)$. For example, using the continuous compounding convention to quote the spot rates $r(t|T)$, we have the relationship

$$B(t|T) = e^{-r(t|T)(T-t)}. \tag{7.1}$$

Thus, by solving for $r(t|T)$ in terms of the bond price $B(t|T)$ as

$$r(t|T) = -\frac{1}{T-t} \ln(B(t)),$$

we see that it is possible to compute the time T spot rate from the price $B(t|T)$. So, if bond prices $B(t|T)$ for various maturity times T are specified, this allows the corresponding spot rates to be computed. If $B(t|T_i)$ is specified for only a finite number of discrete times T_i, then we obtain a discrete term structure whose rates are known only at the discrete times T_i. On the other hand, if $B(t|T)$ is specified for all times $T > t$, then we have a continuous term structure. In both cases, zero-coupon bond prices can be used to compute spot rates and vice versa.

An important interpretation of the bond price $B(t|T)$ is as a *discount factor*, as can be seen from equation (7.1). By a discount factor, we mean that multiplication of a cash flow $X(T)$ at time T by $B(t|T)$ has the effect of converting it into an equivalent cash flow at time t, $X(t) = B(t|T)X(T)$. That is, multiplication by $B(t|T)$ discounts or "moves" a cash flow from time T to t.

Since bonds are readily tradable in markets, zero-coupon bonds will play the role of the marketed tradables in many of our term structure models.

7.2.3 Forward Rates

The third equivalent description of the spot rate curve is through forward interest rates. The concept of a forward interest rate is as follows. Assume the current time is t, and I would like to borrow some money. However, I don't need to borrow the money immediately. Instead, I know that I will need the money a year from now, and that I will be able to repay it two years from now. That is, a year from now, let's call this time T_1, I will need to borrow money from my bank. I will pay back the borrowed money in two years, which we will call time T_2. However, I want to *lock in* the interest rate that I will pay on that loan right now at time t. That locked-in interest rate, which corresponds to borrowing money at time T_1 and repaying at time T_2, is the time t forward rate between T_1 and T_2. We denote this forward interest rate as $f(t|T_1, T_2)$. The cash flow diagram corresponding to this situation is shown in Figure 7.3.

Note the following relationships between forward rates, spot rates, and zero-coupon bonds. The first quite obvious relationship is that the forward rate with $T_1 = t$ is just the spot rate. That is, $f(t|t, T_2) = r(t|T_2)$.

Next, if I have a spot rate to time T_1, $r(t|T_1)$, and a forward rate from time T_1 to time T_2, $f(t|T_1, T_2)$, then I should be able to compute the spot rate to time T_2. This can be done by noting that the forward rate allows me to discount a cash flow from

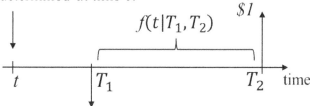

Figure 7.3 Cash flow diagram corresponding to forward rate, $f(t|T_1, T_2)$.

time T_2 to time T_1, and the spot rate $r(t|T_1)$ will then allow me to discount it back to the current time. Since the net effect of these two discounting operations is that a cash flow is discounted from T_2 to t, I should obtain the same result if I had just directly used the spot rate $r(t|T_2)$ for T_2 to discount to time t. Thus, the following discounting relationship should hold if we are using a continuous compounding convention for quoting both spot and forward rates:

$$e^{-r(t|T_2)(T_2-t)} = e^{-r(t|T_1)(T_1-t)} e^{-f(t|T_1,T_2)(T_2-T_1)}.$$

Using this equation to solve for the forward rate $f(t|T_1, T_2)$ in terms of the spot rates gives

$$f(t|T_1, T_2) = \frac{r(t|T_2)(T_2 - t) - r(t|T_1)(T_1 - t)}{T_2 - T_1}. \tag{7.2}$$

Thus, given a spot rate curve, we can use the above relationship to compute the forward rates between any two times. Moreover, given a set of forward rates, $r(t|T_1) = f(t|t, T_1), f(t|T_1, T_2), f(t|T_2, T_3), \ldots, f(t|T_{n-1}, T_n)$, we can compute the spot rates for any of the times T_j, $j = 1 \ldots n$, via

$$
\begin{aligned}
e^{-r(t|T_j)(T_j-t)} &= e^{-f(t|t,T_1)(T_1-t)} e^{-f(t|T_1,T_2)(T_2-T_1)} \ldots e^{-f(t|T_{j-1},T_j)(T_j-T_{j-1})} \\
&= e^{-\sum_{i=1}^{j} f(t|T_{i-1},T_i)(T_i-T_{i-1})}
\end{aligned}
\tag{7.3}
$$

where we have used the convention that $T_0 = t$ in the last equality.

7.2.4 Discount Factors

Not surprisingly, there is a "discount" version of equation (7.3). Recall that the left side is the price of a zero-coupon bond that matures at time T_j, $B(t|T_j)$. As explained previously, we can also interpret $B(t|T_j)$ as a discount factor that discounts cash flows from time T_j to t. In a similar manner, we can view the terms involving the forward rates $e^{-f(t|T_{i-1},T_i)(T_i-T_{i-1})}$ as discount factors that discount cash flows from time T_i to T_{i-1}, but seen from time t. To emphasize this discount interpretation, we can denote these forward discount factor terms as

$$D(t|T_{i-1}, T_i) = e^{-f(t|T_{i-1}, T_i)(T_i - T_{i-1})}. \tag{7.4}$$

See Figure 7.4 for the corresponding cash flow diagram.

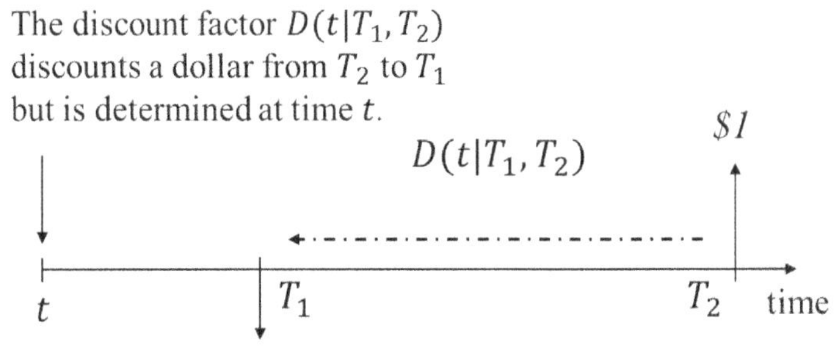

The discount factor $D(t|T_1, T_2)$
discounts a dollar from T_2 to T_1
but is determined at time t.

$D(t|T_1, T_2)$

$1

t T_1 T_2 time

Figure 7.4 Cash flow diagram for a discount factor.

Equation (7.3) can be written in terms of these discount factors as

$$
\begin{aligned}
B(t|T_j) &= B(t|T_1)D(t|T_1, T_2) \cdots D(t|T_{j-1}, T_j) \\
&= \prod_{i=1}^{j} D(t|T_{i-1}, T_i),
\end{aligned}
\tag{7.5}
$$

where we have again used the convention that $T_0 = t$ and that $B(t|T_1) = D(t|T_0, T_1)$.

An alternative interpretation of $D(t|T_{i-1}, T_i)$ is as the forward price of a zero-coupon bond with maturity T_i, $B(t|T_i)$, where the expiration of the forward contract is at time T_{i-1}. This interpretation as forward prices of bonds will be used extensively in our derivation of the LIBOR Market Model in Section 7.5.

The upshot of all this is that spot rates, zero-coupon bonds, forward rates, and discount factors all provide alternative methods of describing the term structure. The relationships in equations (7.3) and (7.5) are shown in Figure 7.5. Before proceeding, let's work an example to make sure that you understand how spot rates, zero-coupon bond prices, forward rates, and discount factors are related.

Example 7.1 (Term Structure)
Let the current time be $t = 0$ and consider the spot rates under a continuous compounding convention for different maturities T_i given in the table below,

Time: T_i	0.1	0.2	0.3	0.4
Spot Rate: $r(0|T_i)$	3.1%	3.2%	3.25%	3.28%.

According to equation (7.1), we can use the spot rates to solve for the zero-coupon

Forward Rates

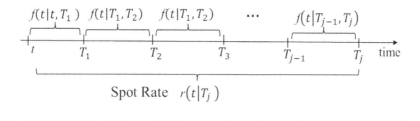

Spot Rate $r(t|T_j)$

Discount Factors

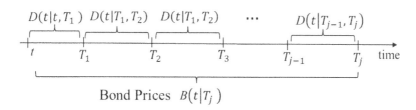

Bond Prices $B(t|T_j)$

Figure 7.5 (Top) Relationship between spot rates and forward rates. (Bottom) Relationship between zero-coupon bond prices and discount factors.

bonds $B(0|T)$. For example,

$$B(0|0.2) = e^{-r(0|0.2)(0.2)} = 0.99362.$$

Additionally, we may compute the forward rates $f(0|T_{i-1}, T_i)$ from equation (7.2). As an example, $f(0|0.2, 0.3)$ may be computed as

$$f(0|0.2, 0, 3) = \frac{r(0|0.3)(0.3) - r(0|0.2)(0.2)}{0.3 - 0.2} = 3.35\%.$$

Lastly, the discount factors between different times can be computed from the forward rates as in equation (7.4). For example, the discount factor between 0.2 and 0.3 is given by

$$D(0|0.2, 0.3) = e^{-f(0|0.2, 0.3)(0.1)} = 0.996656.$$

The following table shows bond prices, forward rates, and discount factors corresponding to the term structure of spot rates:

Time: T_i	0.1	0.2	0.3	0.4
Spot Rate: $r(0|T_i)$	3.1%	3.2%	3.25%	3.28%
Zero-Coupon Bonds: $B(0|T_i)$	0.996905	0.993620	0.990297	0.986966
Forward Rate: $f(0|T_{i-1}, T_i)$	3.10%	3.30%	3.35%	3.37%
Discount Factors: $D(0|T_{i-1}, T_i)$	0.996905	0.996705	0.9966556	0.9966357.

7.2.5 Instantaneous Rates

Above, we considered interest rates that span some finite fixed amount of time, such as between t and T, or T_1 and T_2. It turns out that for various modeling purposes (that we will encounter shortly), it is sometimes convenient to consider interest rates over time spans of an instantaneous dt. The two cases that arise are the instantaneous short rate and instantaneous forward rates.

The instantaneous short rate, which we will denote by $r_0(t)$, is defined as the limit of spot rates as $T \to t$. That is,

$$r_0(t) = \lim_{\Delta t \to 0} r(t|t + \Delta t). \qquad (7.6)$$

The instantaneous short rate process plays a critical role in so-called short rate models of the term structure.

We can consider the same sort of limit for forward rates as well. In this case, we define the instantaneous forward rate using a backward difference convention as

$$f(t|s) = \lim_{\Delta s \to 0} f(t|s - \Delta s, s). \qquad (7.7)$$

The purpose of the backward difference convention is to preserve consistency with other notation that will arise later in the chapter. A forward difference convention that defines $f(t|s) = \lim_{\Delta s \to 0} f(t|s, s + \Delta s)$ would lead to exactly the same results. Consistent with the forward difference convention, we note that the instantaneous short rate is a special case of the instantaneous forward rates with $f(t|t) = r_0(t)$.

Instantaneous forward rates allow us to derive an integral version of equation (7.3) as follows. Let T be a fixed maturity time, and consider taking n equal time steps from t to T of size $\Delta s = \frac{T-t}{n}$. Then, equation (7.3) says that

$$e^{-r(t|T)(T-t)} = e^{-\sum_{i=1}^{n} f(t|t+(i-1)\Delta s, t+i\Delta s)\Delta s}. \qquad (7.8)$$

Finally, taking the limit as n goes to infinity so that $\Delta s \to 0$ turns the sum into an integral, and the forward rates into the instantaneous forward rates. This gives

$$B(t|T) = e^{-r(t|T)(T-t)} = e^{-\int_t^T f(t|s)ds}. \qquad (7.9)$$

This relationship will be used when we consider the Heath–Jarrow–Morton model in Section 7.6 and Appendix 7.9.

7.2.6 Two Choices for Underlying Variables

As stated previously, one of the central goals in interest rate modeling is to create an arbitrage-free model of the term structure. Since we are coming from the factor approach, we need to select factors, underlying variables, and tradables. From above, it should be clear that zero-coupon bonds are a natural choice for tradables. But, what should be chosen as the underlying variables?

In general, it is natural to gravitate toward two different choices for underlying variables that are motivated by our discussion and descriptions above. The first would be to simply choose spot rates corresponding to various maturities as the underlying variables. Surprisingly, this turns out to be difficult to make work. (As an example of some of the problems that arise, see Exercise 7.11.) Thankfully, a related idea does work well. Instead of attempting to use many spot rates as underlying variables, using a single spot rate, the *instantaneous short rate process* as defined in equation (7.6), does the trick. This will lead to the single factor short rate models of Section 7.3.2.

The second choice to use as underlying variables is forward rates. We know from above that forward rates can also be used to describe the term structure, so they are a natural choice as well. When we use forward rates that span a discrete amount of time to model a discrete term structure, this will lead to the so-called LIBOR Market Model, considered in Section 7.5. Surprisingly, starting from a model that uses *discount factors* as the underlying variables and then translating to forward rates is one of the cleanest routes to the results. That route will allow us to take advantage of an analogy with forward contracts.

On the other hand, when we want to model the term structure as a function of a continuous maturity variable T, we will need to use the instantaneous forward rates, as defined in equation (7.7), as underlying variables. This leads to the Heath–Jarrow–Morton model of Section 7.6.

The take-away here is that the main difference between various term structure and interest rate derivative models is whether they use the short rate process or forward rates (either discrete or instantaneous) as underlying variables.

As we traverse the rest of this chapter, at the risk of being redundant, I will attempt to remind you of the key relationships between spot rates, zero-coupon bonds, forward rates, and discount factors. We will start our journey through term structure models and interest rate derivatives with models driven by the instantaneous short rate process as the underlying variable.

7.3 MODELS BASED ON THE SHORT RATE

The first class of models that we consider use the instantaneous short rate process as the underlying variable. The short rate process is key because it drives the rate of return of what we refer to as the *money market account*, which plays the same role as the risk-free bond in equity derivatives. Before jumping into full term structure models, we need to have a clear understanding of how the short rate process relates to this special tradable we call a money market account.

7.3.1 The Short Rate of Interest and a Money Market Account

As a reminder, intuitively speaking, the instantaneous short rate $r_0(t)$ is the interest rate earned from the current time t to time $t + dt$ quoted using continuous compounding. (Note that the notation is similar to the constant risk-free rate r_0 used in the previous chapter. The connection is that they both define the the market price of time λ_0 in their respective settings.)

On a plot of the term structure, the short rate is where the spot rate curve intersects the y-axis as in Figure 7.6. That is, the short rate is the annualized rate

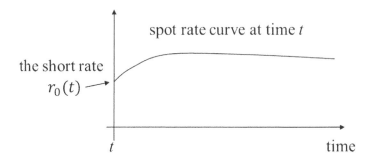

Figure 7.6 Spot rate curve and the short rate process.

of interest that would apply to someone who borrows at the current time t and repays at time $t + dt$.

The short rate $r_0(t)$ is used to describe the instantaneous return on a so-called money market account. We define the *money market account* $B_0(t)$ as the value of an account that is continuously "rolled over" at the instantaneous short rate. That is, you invest your money at the rate $r_0(t)$ over the next dt, then reinvest the resulting amount again over the next dt at the rate $r_0(t + dt)$, and continue in this fashion. Under this idealization of a money market account, the value of the account, $B_0(t)$, follows the dynamics

$$dB_0(t) = r_0(t)B_0(t)dt. \tag{7.10}$$

The money market account plays a special role in interest rate derivatives because its dynamics are *not explicitly driven by a random factor*. Note that this is the case even though the short rate $r_0(t)$ itself can be random and follow a stochastic differential equation of its own. Pay careful attention to this fact in what follows.

In addition to representing the instantaneous rate of interest earned on the money market account, the short rate process is important because in the models we will encounter next, it will be used to drive the *entire* spot rate curve.

7.3.2 Single Factor Short Rate Models

The simplest absence of arbitrage term structure model uses just a single underlying variable—the instantaneous short rate. Furthermore, this variable is driven by a single factor. Hence, models of this type are typically called *single factor short rate models*.

Our goal with these models is to price bonds of various maturities such that there is no arbitrage. Just as we did with equity derivatives, we will approach these problems from the factor model perspective and follow our three step approach.

Step 1: Model and Classify Variables

To start the modeling, we begin with the underlying variable, which is the short rate process,

$$dr_0 = a\,dt + b\,dz, \tag{7.11}$$

where the time dependence is suppressed in $r_0 = r_0(t)$, and $a = a(r_0, t)$ and $b = b(r_0, t)$ can be functions of r_0 and t. Later, we will consider specific choices for $a(r_0, t)$ and $b(r_0, t)$ that define the movement of the short rate. However, for now, when convenient, we will suppress these arguments, so don't always assume that a variable is a constant if no arguments are explicitly listed.

The tradables consist of a money market account, which we denote by B_0, and zero-coupon bonds with face value of \$1 of varying maturities that we denote by $B(t|T)$, where t is the current time and T is the maturity date. The cash flow stream for a zero-coupon bond, $B(t|T)$, was shown in Figure 7.2.

We assume that the zero-coupon bonds are functions of the short rate and time $B(r_0, t|T)$. Then, suppressing arguments so that $B(T) = B(r_0, t|T)$, we can write

$$dB_0 = r_0 B_0 dt, \tag{7.12}$$

$$dB(T) = (B_t(T) + aB_r(T) + \frac{1}{2}b^2 B_{rr}(T))dt + bB_r(T)dz, \tag{7.13}$$

where Ito's lemma was used to derive the equation for $dB(T)$, and B_t, B_r, and B_{rr} represent the partial derivative with respect to t, and the first and second partial derivatives with respect to the short rate r_0, respectively. In particular, when using this partial derivative notation we are also suppressing the subscript on r_0 so that $B_r = B_{r_0}$.

Our classification of variables is

Tradables	Underlying Variables	Factors
$B_0, B(T)$	r_0	dt, dz.

Note that we are not specifically considering an additional derivative security. If you like, you can think of all the bonds $B(T)$ as derivatives, with the short rate being the underlying variable. In any case, any additional derivative will likely look similar to any $B(T)$ since we derived the factor model for $B(T)$ through Ito's lemma.

Now, since $T > t$ can be any time, we actually have an infinite number of tradables. However, since their factor models all "look" the same, we will just write a single equation for $B(T)$ with the implicit assumption that $T > t$ can be any time.

Step 2: Construct a Tradable Table

From the above modeling, we can write the tradable table as

Prices   Value Change Factor Models

$$\begin{bmatrix} B_0 \\ B(T) \end{bmatrix} \quad d\begin{bmatrix} B_0 \\ B(T) \end{bmatrix} = \begin{bmatrix} r_0 B_0 \\ (B_t(T) + aB_r(T) + \frac{1}{2}b^2 B_{rr}(T)) \end{bmatrix} dt + \begin{bmatrix} 0 \\ bB_r(T) \end{bmatrix} dz.$$

Step 3: Apply the Price APT

Finally, we apply and solve the Price APT equation,

$$\left[\begin{array}{c} r_0 B_0 \\ (B_t(T) + a B_r(T) + \frac{1}{2} b^2 B_{rr}(T)) \end{array} \right] = \left[\begin{array}{c} B_0 \\ B(T) \end{array} \right] \lambda_0 + \left[\begin{array}{c} 0 \\ b B_r(T) \end{array} \right] \lambda_1. \quad (7.14)$$

The first equation gives $\lambda_0 = r_0$ which is the random short rate. The second equation leads to

$$B_t(T) + (a - \lambda_1 b) B_r(T) + \frac{1}{2} b^2 B_{rr}(T) = r_0 B(T), \quad (7.15)$$

which is the key absence of arbitrage equation. When applying this equation to zero-coupon bonds, we note that the payoff of each of the bonds is $B(T|T) = 1$.

Note that in models of this form, the market price of risk λ_1 corresponding to dz will likely be *negative*. This is because when the short rate increases, or equivalently the factor dz increases (assuming that the coefficient b is positive in equation (7.11)), the prices of bonds go down. That is, the factor and the tradable move in opposite directions. Thus, for the tradable to be rewarded properly for exposure to the factor risk, the market price of risk λ_1 must be *negative*.

Another point to highlight again is that we did not separate out a derivative from the marketed tradables. In fact, equation (7.15) must hold for zero-coupon bonds of any maturity, and it has been left in terms of the market price of risk λ_1. In fact, since in general we have many bonds, if we assume that λ_1 is constant, any one of them should allow us to solve for λ_1. (Recall that we do not necessarily need to make this assumption. The market prices of risk can be functions of time or any of the information at time t. However, without justification for why the market price of risk would vary, it is reasonable as a first cut to assume that it is a constant.) In practice, different bonds will often give different values of λ_1, indicating that the model is not exactly correct.

When we try to determine a *single* best fit λ_1 from the bond data, we are treating all the bonds as marketed tradables, and this is interpreted as the *calibration* phase as explained in Section 5.6.2. After this calibration phase, we could then use this model to price other interest rate derivatives such as caps, floors, bond options, etc. However, an important use for this model is simply to price bonds in the market with absence of arbitrage, which will be explained next in the context of the Vasicek model.

7.3.2.1 Vasicek Model

Vasicek [51] proposed the following model for the short rate process:

$$dr_0 = \kappa(\theta - r_0)dt + b\,dz. \quad (7.16)$$

Referring to equation (7.11), this corresponds to the specific choices of $a(r_0, t) = \kappa(\theta - r_0)$ and $b(r_0, t) = b$. Hopefully, you will recognize this as an Ornstein–Uhlenbeck process from Section 3.4.1. In this model, the short rate is Gaussian and mean reverting around the level θ. This is a reasonable model for interest rates, since they tend to fluctuate around a fixed level. A drawback of this model is that it allows for the possibility that r_0 can drift negative, implying a negative rate.

With these parameter choices, the absence of arbitrage equation in (7.15) is given by

$$B_t(T) + (\kappa(\theta - r_0) - \lambda_1 b)B_r(T) + \frac{1}{2}b^2 B_{rr}(T) = r_0 B(T), \quad B(T|T) = 1. \quad (7.17)$$

Because the boundary condition is so simple for this problem, it is possible to solve it in closed form. To see how this can be obtained, we first guess a solution of the form

$$B(t|T) = \exp(x(t)r_0(t) + y(t)). \quad (7.18)$$

If we substitute this into (7.17), we obtain

$$(\dot{x}r_0 + \dot{y})B(t|T) + (\kappa(\theta - r_0) - \lambda_1 b)xB(t|T) + \frac{1}{2}b^2 x^2 B(t|T) = r_0 B(t|T). \quad (7.19)$$

Matching constant terms and the coefficients of $r_0(t)$ leads to two ordinary differential equations,

$$\dot{x} = \kappa x + 1, \quad x(T) = 0, \quad (7.20)$$
$$\dot{y} = (-\kappa\theta + \lambda_1 b)x - \frac{1}{2}b^2 x^2, \quad y(T) = 0. \quad (7.21)$$

The first equation is a linear differential equation with solution

$$x(t) = \frac{\exp(\kappa(t - T)) - 1}{\kappa}. \quad (7.22)$$

The second differential equation for $y(t)$ is even simpler to solve. We just integrate after plugging in the solution to $x(t)$. To do this, let's pre-compute the indefinite integrals of $x(t)$ and $x^2(t)$ and use the fact that $\exp(\kappa(t - T)) = 1 + \kappa x(t)$ which follows from equation (7.22) above. We obtain

$$\int x(t)dt = \frac{\exp(\kappa(t - T))}{\kappa^2} - \frac{t - T}{\kappa}$$
$$= \frac{1 + \kappa x(t)}{\kappa^2} - \frac{t - T}{\kappa}$$
$$= \frac{1}{\kappa^2} + \frac{(x(t) - (t - T))}{\kappa}$$

and

$$
\begin{aligned}
\int x^2(t)dt &= \frac{1}{\kappa^2}\left(\frac{\exp(2\kappa(t-T))}{2\kappa} - \frac{2\exp(\kappa(t-T))}{\kappa} + (t-T)\right) \\
&= \frac{1}{\kappa^2}\left(\frac{(1+2\kappa x(t)+\kappa^2 x^2(t))}{2\kappa} - \frac{2(1+\kappa x(t))}{\kappa} + (t-T)\right) \\
&= \frac{1}{2\kappa^3}\left(-2\kappa(x(t)-(t-T)) + \kappa^2 x^2(t) - 3\right) \\
&= \frac{-(x(t)-(t-T))}{\kappa^2} + \frac{x^2(t)}{2\kappa} - \frac{3}{2\kappa^3}.
\end{aligned}
$$

If we integrate equation (7.21), allowing for a constant term $\tilde{K}$ to be determined by the boundary condition $y(T) = 0$, we obtain

$$
\begin{aligned}
y(t) &= (-\kappa\theta + \lambda_1 b)\left(\frac{1}{\kappa^2} + \frac{(x(t)-(t-T))}{\kappa}\right) && (7.23) \\
&\quad -\frac{1}{2}b^2\left(\frac{-(x(t)-(t-T))}{\kappa^2} + \frac{x^2(t)}{2\kappa} - \frac{3}{2\kappa^3}\right) + \tilde{K} && (7.24) \\
&= \left(-\theta + \frac{\lambda_1 b}{\kappa} + \frac{b^2}{2\kappa^2}\right)(x(t)-(t-T)) - \frac{b^2 x^2(t)}{4\kappa}. && (7.25)
\end{aligned}
$$

These equations taken together give the Vasicek closed form bond pricing formula

$$
\begin{aligned}
B(t|T) &= \exp(x(t)r_0(t) + y(t)), && (7.26) \\
x(t) &= \frac{\exp(\kappa(t-T)) - 1}{\kappa}, && (7.27) \\
y(t) &= \left(-\theta + \frac{\lambda_1 b}{\kappa} + \frac{b^2}{2\kappa^2}\right)(x(t)-(t-T)) - \frac{b^2 x(t)^2}{4\kappa}. && (7.28)
\end{aligned}
$$

To calibrate this model, we could use bond prices in the market and choose λ_1 so that this model matches them as closely as possible. The fact that we have a closed form solution makes such a calibration procedure much easier.

Example 7.2 (Vasicek Term Structure)
Assume that the short rate process follows the Vasicek model with parameters $\kappa = 0.25$, $\theta = 0.07$, and $b = 0.02$. Thus, the short rate process mean reverts toward $\theta = 7\%$ and has a standard deviation parameter of $b = 2\%$ per year.

Assume that the market price of risk has been determined at $\lambda_1 = -0.2$. (Recall the discussion of why the market price of risk is likely to be negative from Section 7.3.2.) We can then use equations (7.26) through (7.28) to compute zero-coupon bond prices and hence spot rate curves.

To see the shape of the term structure that the Vasicek model produces, Figure 7.7 plots term structures under a continuous compounding convention for the parameter values given above and for initial values of the short rate of $r_0 = 10\%$, $r_0 = 5\%$, and $r_0 = 1\%$.

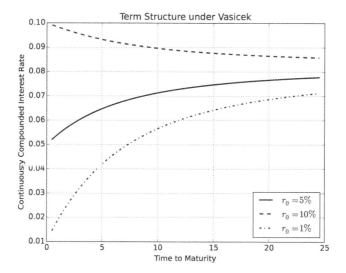

Figure 7.7   Term structures using continuous compounding convention under Vasicek.

7.3.2.2   Cox–Ingersoll–Ross Model

Another popular approach utilizes the Cox–Ingersoll–Ross (CIR) [12] model where the short rate process is assumed to take the form

$$dr_0 = \kappa(\theta - r_0)dt + b\sqrt{r_0}dz. \tag{7.29}$$

This model is sometimes preferred over the Vasicek model because if the parameters in this short rate model are chosen properly (see Section 3.5), the short rate will never go negative. In Vasicek's model, it is possible to have a negative short rate.

Substituting the parameter values defining the short rate process for the CIR model into the absence of arbitrage equation (7.15) leads to

$$(B_t(T) + (\kappa(\theta - r_0) - \lambda_1 b\sqrt{r_0})B_r(T) + \frac{1}{2}b^2 r_0 B_{rr}(T)) = r_0 B(T). \tag{7.30}$$

Now, in order to write a closed form solution for zero-coupon bonds, we will need to massage this equation a bit. First, we will make the assumption that the market price of risk can be written in the form

$$\lambda_1 = \lambda_1' \sqrt{r_0}, \tag{7.31}$$

which then allows the absence of arbitrage equation (7.30) to be expressed as

$$B_t(T) + (\kappa(\theta - r_0) - \lambda_1' b r_0)B_r(T) + \frac{1}{2}b^2 r_0 B_{rr}(T) = r_0 B(T).$$

Next, reparameterizing by equating

$$(\kappa(\theta - r_0) - \lambda_1' b r_0) = \hat{\kappa}(\hat{\theta} - r_0)$$

where

$$\hat{\kappa} = \kappa + \lambda_1' b, \quad \hat{\theta} = \frac{\kappa \theta}{\kappa + \lambda_1' b}$$

leads to

$$B_t(T) + (\hat{\kappa}(\hat{\theta} - r_0))B_r(T) + \frac{1}{2}b^2 r_0 B_{rr}(T) = r_0 B(T).$$

The purpose of these gymnastics is that they allow us to express zero-coupon bonds with payoff $B(T|T) = 1$ through the convenient closed form expression:

$$B(t|T) = Y(t)e^{x(t)r_0(t)}, \tag{7.32}$$

$$\gamma = \sqrt{\hat{\kappa}^2 + 2b^2}, \tag{7.33}$$

$$x(t) = -\frac{2(e^{\gamma(T-t)} - 1)}{(\gamma + \hat{\kappa})(e^{\gamma(T-t)} - 1) + 2\gamma}, \tag{7.34}$$

$$Y(t) = \left[\frac{2\gamma e^{(\hat{\kappa}+\gamma)(T-t)/2}}{(\gamma + \hat{\kappa})(e^{\gamma(T-t)} - 1) + 2\gamma}\right]^{\frac{2\hat{\kappa}\hat{\theta}}{b^2}}. \tag{7.35}$$

7.3.2.3 *Additional Single Factor Short Rate Models*

In addition to the Vasicek and CIR models, there are a number of other single factor short rate models. They differ by how the coefficients $a(r_0, t)$ and $b(r_0, t)$ are chosen in equation (7.11) and go by names such as Rendleman and Bartter [43], Ho-Lee [27], Black-Derman-Toy [3], Hull and White [30], and Black and Karasinski [4]. I will not cover the details of these models here, but rather point the interested readers to the references [28, 36, 38, 33, 53].

7.3.3 Multifactor Short Rate Models

Single factor short rate models are usually not sufficient to describe the term structure well. So, instead of using a single factor model, some authors have proposed multifactor models (see, for example, [10, 35]). They can be approached as follows.

Step 1: Model and Classify Variables
Let X be a vector in $\mathbb{R}^n$ of underlying variables affecting the term structure. We assume that these variables follow a stochastic differential equation model

$$dX = f(X, t)dt + g(X, t)dz, \tag{7.36}$$

where $z \in \mathbb{R}^n$ is a vector of uncorrelated Brownian motions. (Thus, $g(X, t) \in \mathbb{R}^{n \times n}$ and $f(X, t) \in \mathbb{R}^n$.)

We then take the short rate to be a function of the underlying variables X as $r_0(X, t)$. Note that one possibility is to have the short rate be one of the elements in the vector X. Thus, it is possible to choose $r_0(X, t) = X^i$ where X^i is the i-th underlying variable.

To derive the absence of arbitrage equation that zero-coupon bonds would follow, we just note that bonds are functions of X and t and can be represented as $B(X,t|T)$. Via Ito's lemma, our tradable bonds $B(T) = B(X,t|T)$ follow

$$dB(T) = \mathcal{L}B(T)dt + B_r(T)g(X,t)dz$$

where

$$\mathcal{L}B(T) = B_t(T) + B_X(T)f(X,t) + \frac{1}{2}Tr[B_{XX}(T)g(X,t)g^T(X,t)].$$

In addition to these bonds, we also have the money market account,

$$dB_0 = r_0(X,t)B_0dt.$$

In this case, the classification of variables is

Tradables	Underlying Variables	Factors
$B_0, B(T)$	X	$dt, dz.$

Step 2: Construct a Tradable Table
Putting these together in a tradable table gives

$$\begin{array}{c|c} \text{Prices} & \text{Value Change Factor Models} \\ \hline \begin{bmatrix} B_0 \\ B(T) \end{bmatrix} & d\begin{bmatrix} B_0 \\ B(T) \end{bmatrix} = \begin{bmatrix} r_0(X,t)B_0 \\ \mathcal{L}B(T) \end{bmatrix}dt + \begin{bmatrix} 0 \\ B_r(T)g(X,t) \end{bmatrix}dz, \end{array}$$

where I am still suppressing the X and t arguments in $B(T) = B(X,t|T)$.

Step 3: Apply the Price APT
By the Price APT, we have

$$\begin{bmatrix} r_0(X,t)B_0 \\ \mathcal{L}B(T) \end{bmatrix} = \begin{bmatrix} B_0 \\ B(T) \end{bmatrix}\lambda_0 + \begin{bmatrix} 0 \\ B_X(T)g(X,t) \end{bmatrix}\lambda \qquad (7.37)$$

where $\lambda \in \mathbb{R}^n$. The first equation gives $\lambda_0 = r_0(X,t)$ which is the random short rate. The second equation leads to

$$B_t(T) + B_X(T)(f(X,t) - g(X,t)\lambda) + \frac{1}{2}Tr[B_{XX}(T)g(X,t)g^T(X,t)] = r_0(X,t)B(T).$$

For zero-coupon bonds, the boundary condition is $B(X,T|T) = 1$.

An issue that can arise here is the complexity of solving this partial differential equation. If it cannot be solved in a tractable manner, then even calibration of λ to the current term structure can be difficult. It is much easier if the solution to the partial differential equation is known in closed form as a function of λ, as in the case of the single factor Vasicek or CIR models, in which case we can quickly adjust λ to best fit the market data of bonds.

In the next section, we move on to models that use forward rates as underlying variables.

7.4 MODELS BASED ON FORWARD RATES

We just looked at some short rate based models. Next, we consider some different term structure models that use forward rates as underlying variables. This is
actually a natural thing to do when considering interest rate derivatives such as
caplets that can be seen as options on a forward interest rate. We will say a little
more about such derivatives later, but if a fuller understanding is warranted then I
recommend the texts [28, 53]. Before we tackle these models, we need to do some
preliminary work to understand forwards and derivatives on forwards. Trust me,
this preliminary work will be worth the effort.

7.4.1 An Important Detour on Forwards

So why are we taking a detour to talk about forward contracts? Well, as you will see,
some notions in interest rate modeling are just forward contracts. In fact, I believe
using an analogy with forwards is the most natural way to explain the results that
we will see. So, let's spend a little time and understand how to deal with forward
contracts, and this will make life easier when we move to certain forward rate based
interest rate models.

Let's start by considering the forward price $F(t|T)$ of an asset that refers to
the price to be paid at expiration time T (as seen from time t) in exchange for the
asset. Recall that forwards and futures differ because futures contracts are marked
to market whereas forward contracts are not. This affects their pricing (see Exercise
7.13). Here, we are only talking about *forwards*.

Now, we know that there is a simple absence of arbitrage relationship between
the forward price of an asset $F(t|T)$, the spot price of the same asset $S(t)$, and a
zero-coupon bond $B(t|T)$ that pays \$1 at the expiration date of the forward contract
T. A good way to think about $B(t|T)$ is that the price $B(t|T)$ of the bond is simply
a discount factor that discounts a single dollar from time T to time t.

The relationship between these three quantities is $S(t) = B(t|T)F(t|T)$ (see [36]
or [28] for the simple derivation of this). This is depicted in Figure 7.8. In essence,
$B(t|T)$ discounts the forward price back to the present, where it must equal the
spot price, $S(t)$.

We are going to focus our attention on two aspects of dealing with a forward
contract:

1. We will derive a formula for the market price of risk for the factor driving
   the forward price. This will mirror market price of risk relationships in the
   interest rate models that come later.

2. We will change to a new set of units (change of numeraire) to simplify call
   option pricing on forward contracts. This will provide an analogy for caplet
   pricing later.

Spot, Forward, and Bond Relationship

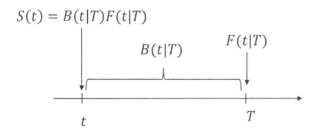

Figure 7.8 Relationship between spot, forward, and bond.

7.4.1.1 Market Price of Risk for Forwards

We will start with Item 1 above and work through our three step procedure with the goal of arriving at the market price of risk associated with the factor driving a forward price.

Let's focus on the basic relationship shown in Figure 7.8 relating the bond $B(t|T)$, the spot price $S(t)$, and the forward price $F(t|T)$:

$$S(t) = B(t|T)F(t|T). \tag{7.38}$$

It is important to point out that $S(t)$ and $B(t|T)$ are tradable, whereas $F(t|T)$ is not. That is, a forward contract is tradable, but a forward price is not. With this in mind, let's proceed with our three step procedure.

Step 1: Model and Classify Variables
Let's start with a simple model of the bond price as

$$dB(t|T) - g_1 B(t|T)dt + h_1 B(t|T)dz_1 \tag{7.39}$$

and the forward price as

$$dF(t|T) = g_2 F(t|T)dt + h_2 F(t|T)dz_2, \tag{7.40}$$

where as usual we allow the possibility that g_1, g_2, h_1, and h_2 are functions with suppressed arguments.

At this point, I must apologize for the seemingly confusing notation choices for the parameters of the bond factor model and the forward price factor model. Note that y_1 and h_1 are associated with the bond that is driven by z_1, and that y_2 and h_2 are associated with the forward price that is driven by z_2. Furthermore, let's assume that z_1 and z_2 are correlated with $\mathbb{E}[z_1 z_2] = \rho dt$. If we are willing to put up with this seemingly confusing notation now, it will actually make our work easier when we move to term structure models.

Now, we know that $S(t) = B(t|T)F(t|T)$ and by Ito's lemma we have

$$
\begin{align}
dS &= d(B(t|T)F(t|T)) \tag{7.41} \\
&= dB(t|T)F(t|T) + B(t|T)dF(t|T) + dB(t|T)dF(t|T) \tag{7.42} \\
&= (g_1 + g_2 + h_1\rho h_2)Sdt + h_1 Sdz_1 + h_2 Sdz_2. \tag{7.43}
\end{align}
$$

When we classify these variables, we emphasize again that $F(t|T)$ is just a forward price and not the value of a forward contract. Thus $F(t|T)$ is not tradable, but rather just an underlying variable, so our classification is

$$
\begin{array}{ccc}
\text{Tradables} & \text{Underlying Variables} & \text{Factors} \\
B(t|T), S(t) & B(t|T), F(t|T) & dt, dz_1, dz_2.
\end{array}
$$

Step 2: Construct a Tradable Table
The tradable table is given by:

$$
\begin{array}{cc}
\text{Prices} & \text{Value Change Factor Models}
\end{array}
$$

$$
\begin{bmatrix} B(t|T) \\ S \end{bmatrix} \quad \bigg| \quad d\begin{bmatrix} B(t|T) \\ S \end{bmatrix} = \begin{bmatrix} g_1 B(t|T) \\ (g_1 + g_2 + h_1\rho h_2)S \end{bmatrix} dt + \begin{bmatrix} h_2 B(t|T) & 0 \\ h_1 S & h_2 S \end{bmatrix} \begin{bmatrix} dz_1 \\ dz_2 \end{bmatrix}.
$$

Note that I don't have a derivative yet, so at this point I am just interested in calibration. That is, I want to determine the market prices of risk λ_1 and λ_2 corresponding to dz_1 and dz_2. Thus, we assume that both $B(t|T)$ and $S(t)$ are marketed. Next, we can move to the Price APT.

Step 3: Apply the Price APT
The Price APT equation reduces to

$$
\begin{bmatrix} g_1 \\ (g_1 + g_2 + h_1\rho h_2) \end{bmatrix} = \lambda_0 \begin{bmatrix} 1 \\ 1 \end{bmatrix} + \begin{bmatrix} h_1 & 0 \\ h_1 & h_2 \end{bmatrix} \begin{bmatrix} \lambda_1 \\ \lambda_2 \end{bmatrix}.
$$

We can simplify this set of equations further by subtracting the first equation from the second, resulting in

$$
\begin{bmatrix} g_1 \\ (g_2 + h_2\rho h_1) \end{bmatrix} = \begin{bmatrix} 1 & h_1 & 0 \\ 0 & 0 & h_2 \end{bmatrix} \begin{bmatrix} \lambda_0 \\ \lambda_1 \\ \lambda_2 \end{bmatrix}. \tag{7.44}
$$

The first equation involving λ_0 and λ_1 is

$$
g_1 = \lambda_0 + h_1\lambda_1. \tag{7.45}
$$

However, we are really after the market price of risk λ_2 for dz_2 which is driving the forward price. In this case, the second equation for λ_2 is

$$
g_2 + h_2\rho h_1 = h_2\lambda_2, \tag{7.46}
$$

and this allows us to solve for the market price of risk λ_2 as

$$\lambda_2 = \frac{g_2 + h_2 \rho h_1}{h_2} = \frac{g_2}{h_2} + h_1 \rho. \tag{7.47}$$

Just for fun, compare this with the result for futures in Section 4.4.2. In that case, translating between the notational differences, the market price of risk was

$$\lambda_2 = \frac{g_2}{h_2}. \tag{7.48}$$

Even under random interest rates, when dealing with futures, we would have obtained this same formula for the market price of risk. Thus, the market price of risk reveals a key distinction between forwards and futures. You are encouraged to explore how this makes a difference in pricing of derivatives on forwards versus futures in Exercise 7.13.

The point of the analysis here was to derive the equation for the market price of risk when dealing with forwards. That is equation (7.46) or (7.47). We will arrive at an exactly analogous formula for market prices of risk when dealing with the LIBOR Market Model (and also use this to explain the Heath–Jarrow–Morton Model). So, please remember the form of the market price of risk in equation (7.47).

To interpret what is going on here, if we look at equation (7.47) we see that the market price of risk is not completely determined by the drift term g_2 and volatility term h_2 of the forward price, as was the case for futures. Instead, when we deal with forwards, the market price of risk is influenced by the correlation of the forward price and the bond price via ρ. The stronger that correlation, the larger the increase in the market price of risk (assuming h_1 is positive). Thus, the key take-away is that the market price of risk for a forward price must refer to the correlation between the forward price and the price of the bond that discounts the forward at time T to the current time t. Let's remember this idea.

Since we did not deal with a derivative security in this section, we can think of the market price of risk equation (7.46) or (7.47) as a calibration equation. Thus, if we have forward prices, we would use equation (7.46) to calibrate and determine market prices of risk that later could be used in pricing some other derivative.

Example 7.3 (Forwards versus Futures)
Consider a forward contract with expiration $T = 0.5$ where the forward price follows the dynamics
$$dF = 0.1F dt + 0.3F dz_2$$
and the zero-coupon bond with maturity $T = 0.5$ follows

$$dB = g_1 B dt + (0.2)(T - t) B dz_1.$$

Assume the correlation between the factors is $\mathbb{E}[dz_1 dz_2] = \rho dt = -0.2 dt$.

Let the current time be $t = 0$. Then according to equation (7.47), the market price of risk λ_2 for dz_2 which is driving the forward price is

$$\lambda_2 = \frac{g_2}{h_2} + h_1\rho = \frac{0.1}{0.3} + (0.2)(0.5 - 0)(-0.2) = 0.3133.$$

On the other hand, if F had been the dynamics describing a *futures* price, the market price of risk λ_2 for dz_2 would have been given by equation (7.48) as,

$$\lambda_2 = \frac{g_2}{h_2} = \frac{0.1}{0.3} = 0.3333.$$

Next, we will assume that we actually have a specific derivative security and want to price it. We will show that in certain cases there is a convenient set of units to use. This insight will be helpful when dealing with specific interest rate derivatives that will arise later in the chapter.

7.4.1.2 Pricing Options on Forwards

This section involves Item 2 on our list in Section 7.4.1. Before showing you the convenient set of units, we will first try it the "hard" way. Hopefully this will convince you of the value of choosing the right units.

Step 1: Model and Classify Variables
Consider actually pricing a derivative on a forward price. For example, we might consider pricing a call option on a forward. Let's assume this call option is a function of the forward price $F(t|T)$ and the short rate r_0 that follows

$$dr_0 = a_0 dt + b_0 dz_1. \tag{7.49}$$

Note that I have assumed that the short rate is driven by the same Brownian motion that is driving $B(t|T)$. We have to be very careful about the consistency of our assumptions in a situation like this. At the end of the day, by choosing the right set of units it won't be an issue, but you should always be watching for potential inconsistencies in your models. In any case, let's continue.

Let $c(F, r_0, t)$ denote the option on the forward, and by Ito's lemma we have

$$dc = \mathcal{L}c + b_0 c_r dz_1 + h_2 F c_F dz_2$$

where

$$\mathcal{L}c = \left(c_t + a_0 c_r + g_2 F c_F + \frac{1}{2} b_0^2 c_{rr} + b_0 \rho h_2 F c_{rF} + \frac{1}{2} h_2^2 F^2 c_{FF} \right).$$

We classify the variable as follows,

Tradables	Underlying Variables	Factors
$B(t|T), S(t), c(F, r_0, t)$	$F(t|T), r_0$	$dt, dz_1, dz_2.$

Step 2: Construct a Tradable Table

Instead of writing the entire tradable table, let's just focus on the derivative that we want to price. In fact, we already performed the calibration that we need in the previous section, so the tradable table for the derivative is

$$
\begin{array}{c}
\text{Prices} \\
---- \\
[\,c\,]
\end{array}
\quad
\begin{array}{c}
\text{Value Change Factor Models} \\
-------------- \\
d\,[\,c\,] = [\,\mathcal{L}c\,]\,dt + [\,b_0 c_r \quad h_2 F c_F\,]
\begin{bmatrix} dz_1 \\ dz_2 \end{bmatrix}.
\end{array}
$$

Step 3: Apply the Price APT

Appealing to the Price APT equation for c gives

$$
c_t + a_0 c_r + g_2 F c_F + \frac{1}{2} b_0^2 c_{rr} + b_0 \rho h_2 F c_{rF} + \frac{1}{2} h_2^2 F^2 c_{FF} = r_0 + \lambda_1 b_0 c_r + \lambda_2 h_2 F c_F,
$$

where we have substituted in for $\mathcal{L}c$ and $r_0 = \lambda_0$. Finally, we would use the market prices of risk derived in equation (7.47) in the previous section for forward prices, leading to

$$
c_t + (a_0 - b_0 \lambda_1) c_r + \frac{1}{2} b_0^2 c_{rr} + b_0 \rho h_2 F c_{rF} + \frac{1}{2} h_2^2 F^2 c_{FF} = r_0 + h_2 \rho h_1 F c_F,
$$

where we could also substitute $\lambda_1 = \frac{g_1 - r_0}{h_1}$ from (7.45) if desired. Even without specifying a boundary condition or payoff characteristic of a derivative, we see that we have a pretty complicated partial differential equation to solve. This is the messy way to go. Next, I will show a very clean way to deal with this problem by choosing the right set of units.

7.4.1.3 Simplified Pricing in Units of the Bond

After our attempt above, we might hope for a simpler solution. Indeed, such a solution is possible as detailed here.

Step 1: Model and Classify Variables

It will turn out that switching to a different set of units will make pricing much easier. Here is the key insight. Using the absence of arbitrage relationship in equation (7.38) between the forward price, spot price, and zero-coupon bond, the forward price can be written as

$$
F(t|T) = \frac{S(t)}{B(t|T)}. \tag{7.50}
$$

Why is this important? Because it says that the forward price $F(t|T)$ is actually the price of the tradable $S(t)$, but expressed in units of the bond $B(t|T)$. If you don't see this, it might be a good time to go back and re-read Section 6.7.2 on exchange one asset for another. If you like, we can then write

$$
F(t|T) = S^B(t) = \frac{S(t)}{B(t|T)}, \tag{7.51}
$$

which is notation indicating that the price of the spot in terms of units of the bond is just the forward price. In this case, the forward price is the price of a tradable in a different set of units! Recall that the fancy way to say that we are expressing quantities in units of the bond is to say that we have selected $B(t|T)$ as the numeraire.

We also note that in this set of units, the zero-coupon bond of maturity T, that is $B(t|T)$, becomes the new risk-free asset with zero risk-free rate:

$$B^B(t|T) = \frac{B(t|T)}{B(t|T)} = 1, \quad t \leq T. \tag{7.52}$$

From this, we see that it is reasonable to assume that an option whose payoff only depends on the forward price $F(t|T)$ would just be a function of $F(t|T)$ and t, as $c^B(F,t)$. (We must always check that these assumptions are correct at the end!). Ito's lemma leads to

$$dc^B = \left(c_t^B + g_2 F c_B^B + \frac{1}{2} h_2^2 F^2 c_{FF}^B \right) dt + h_2 F c_F^B dz_2.$$

In our new set of units, these variables are classified as

Tradables	Underlying Variables	Factors
$B^B(t|T) = 1, F = S^B(t), c^B(F,t)$	$F(t|T)$	dt, dz_2.

Note that $F = S^B$ is now the value of a tradable!

Step 2: Construct a Tradable Table
The above modeling and classification leads to the tradable table:

Prices   Value Change Factor Models

$$\begin{bmatrix} B^B = 1 \\ S^B = F \\ \hline c^B(F,t) \end{bmatrix} \quad d\begin{bmatrix} B^B \\ F \\ \hline c^B(F,t) \end{bmatrix} = \begin{bmatrix} 0 \\ g_2 F \\ \hline c_t^B + g_2 F c_F^B + \frac{1}{2} h_2^2 F^2 c_{FF}^B \end{bmatrix} dt + \begin{bmatrix} 0 \\ h_2 F \\ \hline h_2 F c_F^B \end{bmatrix} dz_2.$$

Step 3: Apply the Price APT
Moving to the Price APT results in

$$\begin{bmatrix} 0 \\ g_2 F \\ \hline c_t^B + g_2 F c_F^B + \frac{1}{2} h_2^2 F^2 c_{FF}^B \end{bmatrix} = \begin{bmatrix} 1 \\ F \\ \hline c^B \end{bmatrix} \lambda_0 + \begin{bmatrix} 0 \\ h_2 F \\ \hline h_2 F c_F^B \end{bmatrix} \lambda_2.$$

Solving for λ_0 and λ_2 using the marketed tradables in the first two equations gives

$$\lambda_0 = 0, \quad \lambda_2 = \frac{g_2}{h_2}. \tag{7.53}$$

Substitution into the third equations leads to

$$c_t^B + g_2 F c_F^B + \frac{1}{2} h_2^2 F^2 c_{FF}^B = \frac{g_2}{h_2} h_2 F c_F^B \tag{7.54}$$

or

$$c_t^B + \frac{1}{2} h_2^2 F^2 c_{FF}^B = 0, \tag{7.55}$$

which is the simplest form of a Black–Scholes partial differential equation when h_2 is constant (compare with equation (6.6)). With the boundary condition for a call option on the forward $c^B(F(T|T), T) = \frac{c(F(T|T),T)}{B(T|T)} = \frac{\max\{F(T|T)-K,\ 0\}}{B(T|T)} = \max\{F(T|T) - K,\ 0\}$ since $B(T|T) = 1$, the solution in units of the bond $B(t|T)$ is just the Black–Scholes formula,

$$c^B(F(t|T), t) = F(t|T)\Phi(d_1) - K\Phi(d_2) \tag{7.56}$$

with

$$d_1 = \frac{\ln(F(t|T)/K) + (\frac{1}{2}h_2^2)(T - t)}{h_2\sqrt{T - t}}, \tag{7.57}$$

$$d_2 = d_1 - h_2\sqrt{T - t}, \tag{7.58}$$

and we are almost done. Remember that we are still in units of the bond. So, to convert the price of the option from units of the bond $B(t|T)$ back to dollars, we multiply the result above by the price of $B(t|T)$ in dollars, which is just $B(t|T)$. This gives the final formula:

$$c = B(t|T)c^B = B(t|T)\left(F(t|T)N(d_1) - KN(d_2)\right). \tag{7.59}$$

This formula is often called Black's model due to its first appearance in Black's paper on options on futures contracts [2].

This demonstrates how easy it can be to price certain derivatives in the right set of units. We will use a similar trick when we deal with caplets which can be thought of as call options on an interest rate. There will be a slight difference because here we are dealing with the forward price and there we will be dealing with forward rates. However, the same ideas will hold.

Example 7.4 (Pricing an Option on a Forward Contract)
Consider a European call option on a forward contract for crude oil with strike price $K = \$50$ and expiration $T = 0.25$. Assume that the current forward price is $F(0|T) = \$55$ and the yearly volatility of the forward price is 30%. The price of a zero-coupon bond that matures at time T is $B(0|T) = 0.9900$.

To compute the value of the option at $t = 0$, we first calculate

$$d_1 = \frac{\ln(55.00/50.00) + (\frac{1}{2}0.3^2)(0.25)}{0.3\sqrt{0.25}} = 0.7104,$$

$$d_2 = 0.7104 - 0.3\sqrt{0.25} = 0.5604.$$

Substituting these values into equation (7.59) gives

$$c = (0.9900)\,(55.00\Phi(0.7104) - 50.00\Phi(0.5604)) = 6.1876.$$

This detour we have taken provides the road map for our presentation of the LIBOR Market Model next. I am sure you will want to refer back to this forward section as we proceed. If you understand the results that we have obtained so far, then you will breeze through the LIBOR Market Model, and its big brother, the Heath–Jarrow–Morton model.

7.5 LIBOR MARKET MODEL

The LIBOR Market Model (LMM) [9, 41] is a term structure model that is particularly useful for dealing with certain derivatives on interest rates. If you are considering a derivative on an interest rate, then it is natural to want to consider interest rates to be the underlying variables. This is what the LMM does. However, to understand the LMM, it will make more sense to first focus on discount factors (which are related to forward interest rates) as underlying variables. Using discount factors allows us to leverage the results from the previous section on forwards. The key connection is that discount factors can be thought of in the context of forward contracts, as described next.

7.5.1 Discount Factors are Forward Prices

Before we jump into the LIBOR Market Model, let's understand why we just did a detour on forwards. The key connection is that discount factors that move a cash flow from one time in the future to another time in the nearer future can be thought of in the context of forward contracts. Consider discounting a dollar from time T_2 to time T_1, where $T_1 < T_2$. Let $D(t|T_1, T_2)$ denote this discount factor as seen at time t. Refer to Figure 7.4 of Section 7.2.4.

$D(t|T_1, T_2)$ should be thought of as follows. I am sitting at time t and I observe the current term structure of interest rates. If I were to borrow an amount at time T_1 and pay back exactly \$1 at time T_2, how much would I be able to borrow at T_1? This value is $D(t|T_1, T_2)$. Said in another way, if I were to move the cash flow of \$1 at time T_2 back to time T_1, taking into account the current term structure of interest rates at time t, how much would it be equivalent to at time T_1? That is, if I would like to "discount" a \$1 cash flow received at time T_2 back to time T_1 in a manner consistent with the interest rate term structure at time t, the discounted value that I would receive at time T_1 would be $D(t|T_1, T_2)$.

For a moment, let's consider a different but related question involving discounting. What if I am sitting at time t, and I would like to discount the \$1 at time T_2 all the way back to time t. Note that I have two options for doing this. I can discount it directly to time t, which just corresponds to the price of a zero-coupon bond at time t that pays a dollar at time T_2. This is denoted by $B(t|T_2)$. On the other hand, I can discount from time T_2 to time T_1 using the discount factor, and then discount from time T_1 to time t using the zero-coupon bond that matures at time T_1, which is $B(t|T_1)$. To be consistent, we must have

$$B(t|T_2) = B(t|T_1)D(t|T_1, T_2). \qquad (7.60)$$

From this relationship, we see that $D(t|T_t, T_2)$ is in essence the forward price of a bond that pays \$1 at time T_2, where the maturity of the forward contract is time T_1. That is, we could write $F(t|T_1) = D(t|T_1, T_2)$ where the spot price of the asset that the forward contract is on is $S(t) = B(t|T_2)$. To be more explicit, we are connecting the equations

$$S(t) = B(t|T_1)F(t|T_1) \leftrightarrow B(t|T_2) = B(t|T_1)D(t|T_1, T_2) \qquad (7.61)$$

where $B(t|T_2)$ plays the role of the spot $S(t)$ and $D(t|T_1, T_2)$ plays the role of the forward $F(t|T_1)$. This spot, forward, and bond relationship of discount factors is shown in Figure 7.9. Once we make this connection and see past the notation differences, what we consider next will follow in a very straightforward manner from the previous section on forwards.

Discount Factors and Bond Relationship

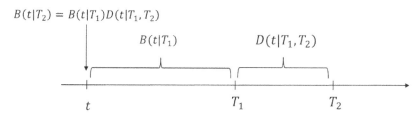

Figure 7.9 Discount factors and the spot, forward, and bond relationship.

7.5.2 LMM with Discount Factors as Underlying Variables

I will begin with a description of the LMM that uses discount factors as the underlying variables to model the term structure of interest rates. This is not the standard description of the LMM, which uses forward rates instead of discount factors. However, later we will convert it to the forward rate description. For now, we will use the discount factors as our underlying variables. A picture of this is given in Figure 7.10. You should contrast this with the short rate models seen in previous sections, and depicted in Figure 7.6, that used the short rate process as the underlying variable to model the term structure.

The LMM model described here is based on discrete discount factors spanning finite periods of time. These times should naturally correspond to the time between interest rates that appear in securities under consideration. For example, if interest/coupon payments on a bond (a floating rate bond, for example) occur each half year and are tied to half-year interest rates, then the discrete times would be each half year and the discount factors would be between those half-year intervals. Taken together, those discount factors could be used to completely characterize the half-year discretized term structure of interest rates, as shown in Figure 7.10.

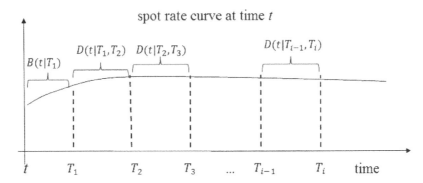

Figure 7.10 LMM using discount factors as underlying variables to describe the term structure of interest rates.

To be specific, let's consider discrete times $t = T_0 < T_1 < T_2 < T_3 < \cdots < T_i \ldots$ and discount factors $D_2 = D(t|T_1, T_2)$, $D_3 = D(t|T_2, T_3), \ldots, D_i = D(t|T_{i-1}, T_i), \ldots$. Note that in our notation we are using the final time of the discount factor as the subscript. For notational convenience, we will also denote the bond that pays \$1 at time T_1 as the first discount factor D_1:

$$D_1 = D(t|t, T_1) = B(t|T_1) = B_1. \tag{7.62}$$

This discount factor is special because it denotes the price of the tradable $B(t|T_1)$. Furthermore, from our discussion above, all the other discount factors can be thought of as forward prices. Let's try to keep that in mind because it will make what we derive next very natural.

For the LIBOR Market Model we will consider the same two items that we considered for forwards, which we enumerate once again.

1. We will derive the formula for the market prices of risk corresponding to the factors that drive the forward rate processes (or equivalently the discount factors).

2. We will show that changing to a new set of units will allow us to easily price certain interest rate derivatives known as caplets.

Let's tackle Item 1 and the market price of risk calibration question first.

7.5.2.1 Calibration of the Market Prices of Risk

Because we did all that background work on forwards and know that discount factors are just forwards, we can almost immediately jump to the answer. Let's apply our market price of risk relationship for forwards from equation (7.46) and see what happens. Our model for B_1 is

$$dB_1 = g_1 B_1 dt + h_1 B_1 dz_1 \qquad (7.63)$$

and recall again that for notational simplicity, we will use alternative notation for the bond as $D_1 = B_1$, so that

$$dD_1 = g_1 D_1 dt + h_1 D_1 dz_1. \qquad (7.64)$$

For the other discount factors, we take

$$dD_i = g_i D_i dt + h_i D_i dz_i \qquad (7.65)$$

for $i > 1$, where the possible dependence of the g's and h's on time or other arguments is being suppressed for notational convenience.

To start simple, we can just deal with B_1, B_2 and D_2 and attempt to determine the market price of risk equation for λ_2 with respect to dz_2 that corresponds to D_2. In this case, our "forward" relationship is

$$B_2 = B(t|T_2) = B(t|T_1)D(t|T_1, T_2) = B_1 D_2. \qquad (7.66)$$

Using the market price of risk equation for forwards given in (7.46), we immediately have

$$h_2 \lambda_2 = g_2 + h_2 \rho_{21} h_1, \qquad (7.67)$$

where ρ_{21} is the correlation coefficient between z_2 and z_1. This is the market price of risk calibration equation for λ_2 relative to the parameters in the discount factor D_2. That was simple, right?

Our goal is to extend this market price of risk calibration to all the discount factors. This follows fairly directly by analogy from the forward market price of risk relationship. First I will present the idea, then we will derive it carefully using our favorite three step procedure.

7.5.2.2 Using the Forward Analogy

Let's consider the "forward" relationship for $D_3 = D(t|T_2, T_3)$. In this case it is

$$B_3 = B(t|T_3) = B(t|T_2)D(t|T_2, T_3) = B_2 D_3. \qquad (7.68)$$

Now, if we go to the market price of risk relationship (7.46), we get a little tripped up because in this case $B_2 = B(t|T_2)$ depends on more than a single Brownian factor. Why is this? Well, from the previous section, $B(t|T_2) = B(t|T_1)D(t|T_1, T_2)$, so that via Ito's lemma

$$dB_2 = (g_1 + g_2 + \rho_{12} h_1 h_2)B_2 dt + h_1 B_2 dz_1 + h_2 B_2 dz_2 \qquad (7.69)$$

which depends on dz_1 and dz_2. However, the market price of risk calibration equation is the natural extension of the case in which the bond depends only on a single factor. That is, the calibration equation is

$$h_3 \lambda_3 = g_3 + h_3 \left(\sum_{k=1}^{2} \rho_{3k} h_k \right) \tag{7.70}$$

where ρ_{3k} is the correlation coefficient between dz_3 and dz_k, and the term $h_3 \left(\sum_{k=1}^{2} \rho_{3k} h_k \right)$ embodies the correlation of the discount factor D_3 (our "forward") with the bond discounting it (B_2). That was the interpretation of this term in the forward case that we are using for motivation. More generally, this basic calibration equation extends to all the discount factors D_i, $i > 1$ as

$$h_i \lambda_i = g_i + h_i \left(\sum_{k=1}^{i-1} \rho_{ik} h_k \right). \tag{7.71}$$

This is really just the forward market price of risk relationship.

If you do not trust my arguments here, I take the careful route in the next section using our three step procedure. However, don't fool yourself into thinking that this situation is more complicated than it really is. If you understand our simple calculation of the market price of risk for forwards, then you already know what I am doing.

7.5.2.3 Formal Three Step Approach

Let's carefully derive equation (7.71).

Step 1: Model and Classify Variables
Recall our notation that $B_1 = B(t|T_1)$, $B_2 = B(t|T_2)$, etc., and $D_1 = B_1$, $D_2 = D(t|T_1, T_2)$, $D_3 = D(t|T_2, T_3)$, etc.. Now, bonds move \$1 from the maturity date all the way back to time t and are tradable. We can relate the price of a bond to the discount factors, which are the underlying variables, by the relationships,

$$B_1 = D_1 \tag{7.72}$$

$$B_2 = \prod_{j=1}^{2} D_j \tag{7.73}$$

$$\vdots$$

$$B_i = \prod_{j=1}^{i} D_j. \tag{7.74}$$

With this notation, we can also write these equations recursively as

$$B_1 = D_1 \tag{7.75}$$

$$B_2 = B_1 D_2 \tag{7.76}$$

$$\vdots$$

$$B_i = B_{i-1} D_i \tag{7.77}$$

which captures the "spot, bond, forward relationship." Figure 7.11 shows this relationship between bonds and discount factors. We are already very aware of this analogy, but it doesn't hurt to review it before diving into some detailed calculations.

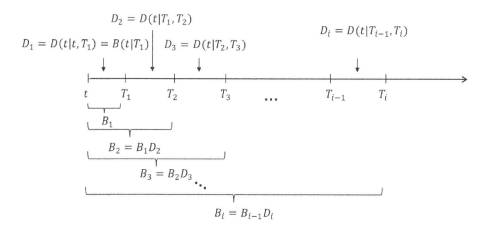

Figure 7.11 Cash flow timing for LMM.

Applying Ito's lemma to equation (7.74), which is a bit involved, results in the following factor model for the tradable bond prices based on the discount factors:

$$
\begin{aligned}
dB_i &= \sum_{j=1}^{i}(g_j dt + h_j dz_j)\prod_{l=1}^{i}D_l + \sum_{j=1}^{i}\sum_{k=1}^{j-1}\rho_{jk}h_j h_k \prod_{l=1}^{i}D_l dt \\
&= B_i\left[\sum_{j=1}^{i}(g_j dt + h_j dz_j) + \sum_{j=1}^{i}\sum_{k=1}^{j-1}\rho_{jk}h_j h_k dt\right] \\
&= B_i\left[\mu_i dt + \sum_{j=1}^{i}h_j dz_j\right],
\end{aligned}
$$

where in the final equation we are defining μ_i as

$$\mu_i = \sum_{j=1}^{i}g_j + \sum_{j=1}^{i}\sum_{k=1}^{j-1}\rho_{jk}h_j h_k. \tag{7.78}$$

The variables in this model are classified as

$$
\begin{array}{ccc}
\text{Tradables} & \text{Underlying Variables} & \text{Factors} \\
B_1, B_2, \ldots, B_i & D_1, D_2, \ldots, D_i & dt, dz_1, dz_2, \ldots, dz_i
\end{array}
$$

which mirrors our classification for spots, bonds and forwards.

Step 2: Construct a Tradable Table

Using the above modeling, we create a tradable table for the bonds, B_i, $i \geq 1$, as

Prices   Value Change Factor Models

$$
\begin{bmatrix} B_1 \\ B_2 \\ \vdots \\ B_i \end{bmatrix}
\quad
d \begin{bmatrix} B_1 \\ B_2 \\ \vdots \\ B_i \end{bmatrix}
=
\begin{bmatrix} B_1 g_1 \\ B_2 \mu_2 \\ \vdots \\ B_i \mu_i \end{bmatrix} dt
+
\begin{bmatrix}
B_1 h_1 & 0 & \ldots & 0 \\
B_2 h_1 & B_2 h_2 & \ldots & 0 \\
& & \ddots & \\
B_i h_1 & B_i h_2 & \ldots & B_i h_i
\end{bmatrix}
\begin{bmatrix} dz_1 \\ dz_2 \\ \vdots \\ dz_i \end{bmatrix}.
$$

Step 3: Apply the Price APT

Imposing the Price APT means that we need to solve the following:

$$
\begin{bmatrix} B_1 g_1 \\ B_2 \mu_2 \\ \vdots \\ B_i \mu_i \end{bmatrix}
=
\begin{bmatrix} B_1 \\ B_2 \\ \vdots \\ B_i \end{bmatrix} \lambda_0
+
\begin{bmatrix}
B_1 h_1 & 0 & \ldots & 0 \\
B_2 h_1 & B_2 h_2 & \ldots & 0 \\
& & \ddots & \\
B_i h_1 & B_i h_2 & \ldots & B_i h_i
\end{bmatrix}
\begin{bmatrix} \lambda_1 \\ \lambda_2 \\ \vdots \\ \lambda_i \end{bmatrix},
$$

which, upon substituting back in for μ_i from equation (7.78), can be reduced to

$$
\begin{bmatrix}
g_1 \\
\left(\sum_{j=1}^{2} g_j + \sum_{j=1}^{2} \sum_{k=1}^{j-1} \rho_{jk} h_j h_k \right) \\
\vdots \\
\left(\sum_{j=1}^{i} g_j + \sum_{j=1}^{i} \sum_{k=1}^{j-1} \rho_{jk} h_j h_k \right)
\end{bmatrix}
=
\begin{bmatrix}
1 & h_1 & 0 & \ldots & 0 \\
1 & h_1 & h_2 & \ldots & 0 \\
& & & \ddots & \\
1 & h_1 & h_2 & \ldots & h_i
\end{bmatrix}
\begin{bmatrix} \lambda_0 \\ \lambda_1 \\ \lambda_2 \\ \vdots \\ \lambda_i \end{bmatrix}.
$$

Finally, we leave the first equation intact and then subtract each previous equation from the next, which reduces this system of equations to

$$
\begin{bmatrix}
g_1 \\
g_2 + h_2 \rho_{21} h_1 \\
\vdots \\
g_i + h_i \sum_{k=1}^{i-1} \rho_{ik} h_k
\end{bmatrix}
=
\begin{bmatrix}
1 & h_1 & 0 & \ldots & 0 \\
0 & 0 & h_2 & \ldots & 0 \\
& & & \ddots & \\
0 & 0 & 0 & \ldots & h_i
\end{bmatrix}
\begin{bmatrix} \lambda_0 \\ \lambda_1 \\ \lambda_2 \\ \vdots \\ \lambda_i \end{bmatrix}.
$$

Note the similarity between this Price APT equation and the one for forwards in (7.44). This is clearly the generalization of that equation. To find the market prices of risk, we need to solve these equations, which can be done quite trivially. From the first equation we obtain

$$g_1 = \lambda_0 + \lambda_1 h_1, \tag{7.79}$$

and from the other equations we have,

$$g_i + h_i \left(\sum_{k=1}^{i-1} \rho_{ik} h_k \right) = h_i \lambda_i, \quad i > 1, \tag{7.80}$$

which is the same result as in equation (7.71). This is the key market price of risk equation for discount factors, and this equation is an important one underlying the LIBOR Market Model. For there to be no arbitrage, there must exist λ's that satisfy these equations.

Here, I used discount factors as the underlying variables. However, the LMM is often written not in terms of the parameters of the discount factors, but rather in terms of parameters describing forward interest rates. Let's see how this works.

7.5.3 Forward Rates as Underlying Variables

We have been doing our calculations using discount factors as the underlying variables. This was depicted in Figure 7.10 and is a useful approach because we can think about discount factors as forwards. However, it is common to deal with forward interest rates as the underlying variables that drive the term structure. This basic setup is depicted in Figure 7.12. As is typical in the LIBOR Market Model, we will use a discrete simple convention for quoting these interest rates (which will be described shortly), rather than the continuous compounding convention of Section 7.2.3. Forward rates under a simple convention will be denoted by $F(t|T_1, T_2)$, as opposed to $f(t|T_1, T_2)$ under a continuous compounding convention.

In this case, we need to translate all our results from discount factors to forward interest rates. After seeing this done, I hope you will appreciate that we began from the discount factor/forward perspective because it allows us to see the big picture more clearly. We will find that the equations in terms of forward interest rates are rather messy and the big picture can easily be lost if we insist on using interest rates as the starting point.

The idea is simple. We need to translate discount factors to forward interest rates $F_i = F(t|T_{i-1}, T_i)$ which are defined relative to the discount factors via

$$D_i = \frac{1}{(1 + \tau_i F_i)} \tag{7.81}$$

where F_i is an annualized forward rate quoted using a simple interest convention, and $\tau_i = T_i - T_{i-1}$ is just a constant that scales the interest rate to the time period of the discount factor. For example, if the discount factor is discounting from time $T_2 = 1$ year to $T_1 = 0.5$ years, and $F_2 = 6\%$, then F_2 is quoted as a yearly interest rate and $\tau_2 = 1 - 0.5 = 0.5$ years is used to scale it to $\tau_2 F_2 = 3\%$ for the half-year time period of the discount factor. See Figure 7.13 for this relationship.

The LIBOR Market Model
Forward Rates as Underlying Variables

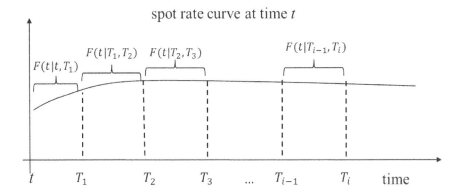

Figure 7.12 LMM using forward rates as the underlying variables to model the term structure.

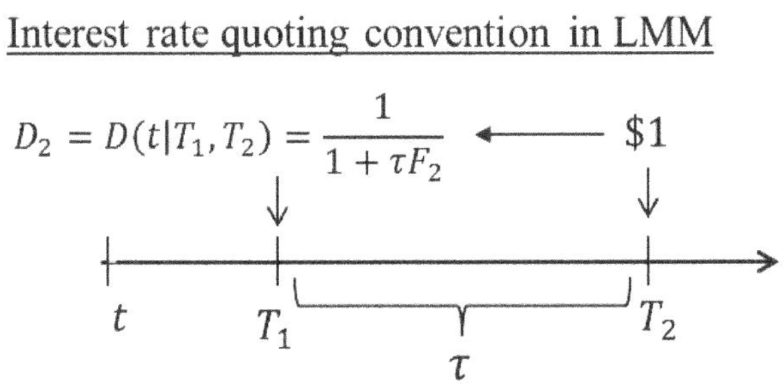

Interest rate quoting convention in LMM

$$D_2 = D(t|T_1, T_2) = \frac{1}{1 + \tau F_2} \longleftarrow \$1$$

Figure 7.13 Simple interest quoting convention relative to the discount factors.

To perform our translation from discount factors to forward interest rates, we assume that D_i is a function of the rates F_i as given in (7.81), and that the forward rates satisfy

$$dF_i = a_i F_i dt + b_i F_i dz_i, \qquad (7.82)$$

where, as usual, a_i and b_i may be hiding time or other dependence.

For the first interest rate F_1 (note that it is actually a spot rate since $R(t|T) = F(t|t, T)$) that corresponds to $D_1 = B_1$, we have

$$B_1 = D_1 = \frac{1}{1 + (T_1 - t)F_1}, \qquad (7.83)$$

so that by Ito's lemma

$$
\begin{aligned}
dD_1 &= \frac{F_1}{(1 + (T_1 - t)F_1)^2}dt + \frac{-(T_1 - t)}{(1 + (T_1 - t)F_1)^2}dF_1 + \frac{1}{2}\frac{2(T_1 - t)^2}{(1 + (T_1 - t)F_1)^3}b_1^2 F_1^2 dt \\
&= \frac{F_1}{(1 + (T_1 - t)F_1)}D_1 dt + \frac{-(T_1 - t)}{(1 + (T_1 - t)F_1)}D_1(a_1 F_1 dt + b_1 F_1 dz_1) \\
&\qquad + \frac{(T_1 - t)^2}{(1 + (T_1 - t)F_1)^2}b_1^2 F_1^2 D_1 dt \\
&= \left(\frac{F_1}{(1 + (T_1 - t)F_1)} + \frac{-(T_1 - t)a_1 F_1}{(1 + (T_1 - t)F_1)} + \frac{(T_1 - t)^2 b_1^2 F_1^2}{(1 + (T_1 - t)F_1)^2} \right) D_1 dt \\
&\qquad + \frac{-(T_1 - t)b_1 F_1}{(1 + (T_1 - t)F_1)}D_1 dz_1 \\
&= \left(\frac{F_1}{(1 + \tau_1 F_1)} + \frac{-\tau_1 a_1 F_1}{(1 + \tau_1 F_1)} + \frac{\tau_1^2 b_1^2 F_1^2}{(1 + \tau_1 F_1)^2} \right) D_1 dt + \frac{-\tau_1 b_1 F_1}{(1 + \tau_1 F_1)}D_1 dz_1,
\end{aligned}
$$

where in the last equality we used the notation $\tau_1 = T_1 - t$. Matching this with the generic stochastic differential equation for D_1 written as

$$dD_1 = g_1 D_1 dt + h_1 D_1 dz_1 \qquad (7.84)$$

gives the correspondence

$$g_1 = \left(\frac{F_1}{(1 + \tau_1 F_1)} + \frac{-\tau_1 a_1 F_1}{(1 + \tau_1 F_1)} + \frac{\tau_1^2 b_1^2 F_1^2}{(1 + \tau_1 F_1)^2} \right), \qquad h_1 = \frac{-\tau_1 b_1 F_1}{(1 + \tau_1 F_1)}. \qquad (7.85)$$

We can perform the same calculation for the other discount factors D_i, $i > 1$. The only difference is that each $\tau_i = (T_i - T_{i-1})$ is a constant and not a function of time t. Ito's lemma gives us

$$
\begin{aligned}
dD_i &= \frac{-\tau_i}{(1 + \tau_i F_i)^2}dF_i + \frac{1}{2}\frac{2\tau_i^2}{(1 + \tau_i F_i)^3}b_i^2 F_i^2 dt \\
&= \frac{-\tau_i}{(1 + \tau_i F_i)}D_i(a_i F_i dt + b_i F_i dz_i) + \frac{\tau_i^2}{(1 + \tau_i F_i)^2}b_i^2 F_i^2 D_i dt \\
&= \left(\frac{-\tau_i a_i F_i}{(1 + \tau_i F_i)} + \frac{\tau_i^2 b_i^2 F_i^2}{(1 + \tau_i F_i)^2} \right) D_i dt + \frac{-\tau_i b_i F_i}{(1 + \tau_i F_i)}D_i dz_i.
\end{aligned}
$$

Comparing this with our previous model for the discount factors for $i > 1$,

$$dD_i = g_i D_i dt + h_i D_i dz_i, \tag{7.86}$$

we identify

$$g_i = \left(\frac{-\tau_i a_i F_i}{(1 + \tau_i F_i)} + \frac{\tau_i^2 b_i^2 F_i^2}{(1 + \tau_i F_i)^2} \right), \quad h_i = \frac{-\tau_i b_i F_i}{(1 + \tau_i F_i)}. \tag{7.87}$$

Plugging this into our market price of risk equation (7.80) for $i > 1$ results in

$$\left(\frac{-\tau_i a_i F_i}{(1 + \tau_i F_i)} + \frac{\tau_i^2 b_i^2 F_i^2}{(1 + \tau_i F_i)^2} \right) + \frac{\tau_i b_i F_i}{(1 + \tau_i F_i)} \left(\sum_{k=1}^{i-1} \frac{\rho_{ik} \tau_k b_k F_k}{(1 + \tau_k F_k)} \right) = \frac{-\tau_i b_i F_i}{(1 + \tau_i F_i)} \lambda_i$$

which can be reduced to

$$a_i - b_i \left(\sum_{k=1}^{i} \frac{\rho_{ik} \tau_i b_k F_k}{(1 + \tau_i F_k)} \right) = b_i \lambda_i, \quad i > 1. \tag{7.88}$$

This is the LMM market price of risk calibration equation. It may look complicated, but we know better. It is just the market price of risk equation for forwards!

This equation is important because it allows us to observe forward interest rates F_i, model them via the stochastic differential equation (7.82), and calibrate the market prices of risk. Compare this to the short rate based models. In that case we were using the instantaneous short rate process $r_0(t)$ as the underlying variable, which is, for practical purposes, unobservable. Yet, we still assumed that we had a model on which to base our pricing. In this case, we really can observe the forward rates F_i and thus it is much easier to imagine developing a reasonable model for them upon which to base pricing.

Here, we have presented only the calibration equations. If we desire to price a derivative that depends on the underlying variables, we would have to write the absence of arbitrage equation for the derivative using the market prices of risk as calibrated above, and solve the resulting equation subject to the appropriate payoff characteristics of the derivative. Depending on the derivative, this can be a daunting task, and I won't pursue this path here. However, we will consider the related task of numerically computing solutions in Chapter 9. To be honest, solving the absence of arbitrage equation is often done using Monte Carlo methods that align better with notions of risk neutral pricing that will be introduced in Chapter 10.

While pricing a general derivative may be tricky, when we switch to the right set of units we can arrive at a very simple pricing formula for certain options. In this case we will consider Item 2 on our list in Section 7.5.2, and tackle the specific problem of pricing an interest rate caplet. Surprisingly, we arrive at a simple closed form solution related to Black's model of Chapter 6, Section 6.5.1. Moreover, once you know how to price a caplet, then you can extend the basic idea to many other popular interest rate derivatives such as caps, floorlets, floors, swaptions, etc. These formulas are a powerful consequence of the LMM, and a good reason for its popularity.

7.5.4 Pricing a Caplet in the Right Units

This is Item 2 on our list from the end of Section 7.5.2. A caplet is a derivative contract that allows an interest payment at a future time to be capped. Consider the following scenario. At time T_{i-1}, the interest rate $F_i = F(T_{i-1}|T_{i-1}, T_i)$ is used to specify an interest payment that will occur at time T_i on a loan amount of P. Specifically, if $\tau = T_i - T_{i-1}$ is the time between interest payments, or the so-called *natural time lag*, then the interest amount that must be paid at time T_i is $PF_i\tau$. That is, since F_i is an annualized interest rate, $F_i\tau$ is the interest rate over the time period τ, and thus $PF_i\tau$ is the total payment amount. The timing of these cash flows is shown in Figure 7.14.

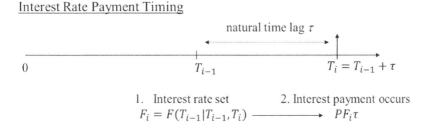

Figure 7.14 Timing of interest payment.

Let K represent the level at which the interest rate F_i can be capped. This ability to cap the interest rate on a single interest payment is called a *caplet*. The interest rate caplet allows the holder to pay only $PK\tau$ instead of $PF_i\tau$ if $F_i > K$. Therefore, we can view the payoff of the caplet as:

$$c(F_i, T_i) = \tau P \max\{F_i - K, 0\}. \qquad (7.89)$$

This looks similar to a standard European call option where F_i takes the place of the stock price and K is the strike price. However, there is a subtle difference. The interest rate $F_i = F(T_{i-1}|T_{i-1}, T_i)$ is known at time T_{i-1} and at that time the caplet option is either exercised or discarded. But, the interest payment, and hence the payoff of the caplet occurs at time T_i. That is, there is a "mismatch" between the exercise time of the option, which is T_{i-1}, and the payoff time T_i.

One way to remedy this mismatch is to discount the payoff at time T_i back to time T_{i-1}. That is, at time T_{i-1}, the payoff is known and becomes a deterministic cash flow at time T_i. Thus, we can discount the payoff cash flow from time T_i back to time T_{i-1} using the zero-coupon bond that matures at time T_i. The payoff cash flow of $P\tau \max\{F_i - K, 0\}$ at time T_i discounted back to time T_{i-1} becomes

$$c(F_i, T_{i-1}) = B(T_{i-1}|T_i)P\tau \max\{F_i - K, 0\} = B_iP\tau \max\{F_i - K, 0\}. \qquad (7.90)$$

This discounting removes the "mismatch" in the timing of the exercise decision and the payoff. This is shown in Figure 7.15.

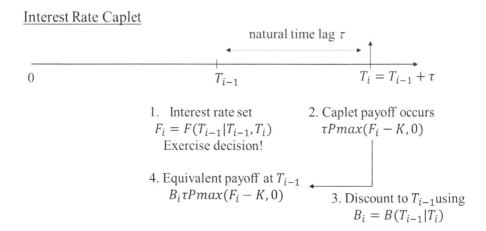

Interest Rate Caplet

Figure 7.15 Timing of caplet payoff cash flows.

Notwithstanding the subtle difference just mentioned, the caplet looks like a call option on a forward interest rate F_i. We saw previously in Section 7.4.1.3 that when dealing with forwards, changing to the right set of units made all the difference. Once again, you will be shocked to see how clean and simple things become when we change to this right set of units. Here, we are dealing with forward interest rates, and not a forward price, but it turns out that the distinction is minor in terms of approach. Let's tackle it using our three step approach.

Step 1: Model and Classify Variables
The key recognition is that the forward interest rate $F_i = F(t|T_{i-1}, T_i)$ can be written as

$$F_i = \frac{1}{\tau}\left(\frac{1}{D_i} - 1\right) = \frac{1}{\tau}\left(\frac{B_{i-1}}{B_i} - 1\right) = \frac{1}{\tau}\left(\frac{B_{i-1} - B_i}{B_i}\right), \qquad (7.91)$$

where recall our notation that $D_i = D(t|T_{i-1}, T_i)$, $B_i = B(t|T_i)$, and $B_{i-1} = B(t|T_{i-1})$. This comes from rearranging equation (7.81) and using (7.77).

Why is this equation so important? If you recall previously when we were dealing with forward prices, we recognized that a forward price was actually the price of a tradable when measured in units of the bond; recall equation (7.51).

In this case, at least at first glance, it is not so obvious that the forward rate F_i is actually the value of something tradable when measured in a different set of units. But upon careful examination, we note that F_i is indeed the price of a tradable. In particular, it is a portfolio that is long $\frac{1}{\tau}$ of B_{i-1} and short $\frac{1}{\tau}$ of B_i, but measured in units of B_i. That is, the denominator of equation (7.91) is the bond B_i, and thus we are denominating in units of this particular bond, which is our numeraire asset. Moreover, the numerator is $\frac{1}{\tau}(B_{i-1} - B_i)$ which is just a long position of $\frac{1}{\tau}$ in bond B_{i-1} and a short position of $\frac{1}{\tau}$ in bond B_i. This is the value of a tradable portfolio.

Note what we have done here. We have treated F_i, which can be associated with a portfolio of bonds measured in units of B_i, as a tradable. That is, we can certainly buy and sell this portfolio as a package, and thus we will argue that it is just as valid a tradable as any other "single asset" tradable. Hence, it must satisfy the Price APT equation when our "currency" is the bond B_i. This recognition will simplify our analysis immensely, and is analogous to the way we treated the forward price when dealing with forward contracts.

Turning our attention to the caplet, we recognize from the payoff equation (7.90) that if we divide the caplet payoff by the numeraire B_i, the payoff of the caplet takes the simple form

$$c^B(F_i, T_{i-1}) = \frac{c(F_i, T_{i-1})}{B_i} = P\tau \max\{F_i - K, 0\}, \qquad (7.92)$$

where the superscript B denotes that we are in units of the bond B_i. From this it seems reasonable to assume that in units of the bond, the caplet will be representable as a function of F_i and t as $c^B(F_i, t)$. Using our factor model of F_i from equation (7.82) and applying Ito's lemma leads to

$$dc^B = \left(c_t^B + a_i F_i c_F^B + \frac{1}{2} b_i^2 F_i^2 c_{FF}^B\right) dt + b_i F_i c_F^B dz_i.$$

Finally, the variables are classified as

$$
\begin{array}{ccc}
\text{Tradables} & \text{Underlying Variables} & \text{Factors} \\
B_i^B = 1, F_i, c^B(F_i, t) & F_i & dt, dz_i
\end{array}
$$

where, as just mentioned above, the superscript B indicates we are expressing quantities in units of the bond B_i.

Step 2: Construct a Tradable Table
The tradable table is

$$
\begin{array}{cc}
\text{Prices} & \text{Value Change Factor Models}
\end{array}
$$

$$
\begin{bmatrix} B_i^B \\ F_i \\ \hline c^B(F_i, t) \end{bmatrix}
\quad
d\begin{bmatrix} B_i^B \\ F_i \\ \hline c^B(F_i, t) \end{bmatrix}
-
\begin{bmatrix} 0 \\ a_i F_i \\ \hline c_t^B + a_i F_i c_F^B + \frac{1}{2} b_i^2 F_i^2 c_{FF}^B \end{bmatrix} dt
+
\begin{bmatrix} 0 \\ b_i F_i \\ \hline b_i F_i c_F^B \end{bmatrix} dz_i.
$$

Step 3: Apply the Price APT

Moving to the Price APT yields

$$
\begin{bmatrix} 0 \\ a_i F_i \\ -- \\ c_t^B + a_i F_i c_F^B + \frac{1}{2} b_i^2 F_i^2 c_{FF}^B \end{bmatrix} = \begin{bmatrix} B_i^B \\ F_i \\ -- \\ c^B \end{bmatrix} \lambda_0 + \begin{bmatrix} 0 \\ b_i F_i \\ -- \\ b_i F_i c_F^B \end{bmatrix} \lambda_i.
$$

Solving using the marketed tradables gives

$$
\lambda_0 = 0, \quad \lambda_i = \frac{a_i}{b_i}, \tag{7.93}
$$

and substituting into the third equation for our derivative leads to

$$
c_t^B + a_i F_i c_F^B + \frac{1}{2} b_i^2 F_i^2 c_{FF}^B = b_i F_i c_F^B \frac{a_i}{b_i}. \tag{7.94}
$$

Upon rearranging, we have

$$
c_t^B + \frac{1}{2} b_i^2 F_i^2 c_{FF}^B = 0, \tag{7.95}
$$

which is the simplest form of the Black–Scholes equation if b_i is constant.

7.5.4.1 LMM Caplet Formula

We must solve the above partial differential equation (7.95) using the boundary condition corresponding to the payoff of the caplet in units of the bond. This was given in equation (7.92) as

$$
c^B(F_i, T_{i-1}) = \frac{B_i P \tau \max\{F_i - K, 0\}}{B_i} = P\tau \max\{F_i - K, 0\}.
$$

This is the standard form of the payoff of a European call option on F_i with strike K. Thus, assuming the volatility parameter b_i of the forward rate F_i is constant, we have a closed form solution in units of the bond,

$$
c^B(F_i, t) = P\tau \left(F_i \Phi(d_1) - K\Phi(d_2) \right), \tag{7.96}
$$

with

$$
d_1 = \frac{\ln(F_i/K) + (\frac{1}{2} b_i^2)(T_{i-1} - t)}{b_i \sqrt{T_{i-1} - t}}, \tag{7.97}
$$

$$
d_2 = d_1 - b_i \sqrt{T_{i-1} - t}, \tag{7.98}
$$

and we are almost done! We just have to change units back to dollars to arrive at our final answer,

$$
c(F_i, t) = B_i c^B(F_i, t) = B(t|T_i) P\tau \left(F_i \Phi(d_1) - K\Phi(d_2) \right). \tag{7.99}
$$

This is a shockingly simple formula, and one of the main reasons the LMM is so popular. This formula is sometimes also referred to as Black's model in reference to his 1976 paper [2] in which a similar formula is presented in the context of options on futures.

Example 7.5 (Pricing a Caplet under LMM)

Consider a caplet that matures at $T = 1$ year with a strike rate of $K = 3\%$. Assume the caplet refers to the 6-month interest rate (i.e., $\tau = 0.5$) and is on a principal amount of $100 million. Let the current forward rate be $F_{1.5} = F(0|1, 1.5) = 2.9\%$, and the volatility of the forward rate be $b_{1.5} = 10\%$. Finally, assume that the zero-coupon bond maturing in 1.5 years is selling for $B_{1.5} = 0.96$.

To compute the absence of arbitrage price of the caplet, we refer to the formula given in equations (7.99) above. From the given information, we note that $T_{i-1} = 1$, $T_i = T_{i-1} + \tau = 1.5$, and d_1 and d_2 can be computed from (7.97) and (7.98) as

$$d_1 = \frac{\ln(0.029/0.03) + (0.5)(0.1)^2(1)}{(0.1)\sqrt{1}} = -0.289,$$

$$d_2 = d_1 - (0.1)\sqrt{1} = -0.389.$$

Substituting these numbers into the LMM formula (7.99) for a caplet gives

$$c = (0.96)(100,000,000)(0.5)\left(0.029\Phi(-0.289) - 0.03\Phi(-0.389)\right) = \$35,677.81.$$

Just as a caplet resembles a call option, a so-called *floorlet*, that we denote here by $f(F_i, t)$, is analogous to a put option. It places a floor or lower limit on the interest rate corresponding to a single interest payment and has the payoff function

$$f(F_i, T_{i-1}) = B_i P \tau \max\{K - F_i, 0\},$$

where K is again the strike rate. As you might guess, the closed form solution for a floorlet is the "put version" of the caplet formula

$$f(F_i, t) = B(t|T_i)P\tau\left(K\Phi(-d_2) - F_i\Phi(-d_1)\right). \tag{7.100}$$

7.5.4.2 From Caplets to Floorlets and Beyond

We derived the formula for a caplet, and by extension, a floorlet. Moreover, these formulas can be built upon to price more complicated interest rate derivatives. For example, a *cap* places an upper limit on all the interest payments corresponding to a loan, while a *floor* imposes a lower limit. Caps and floors can be simply priced as portfolios of caplets and floorlets, respectively. That is, the value of a cap is just the sum of the prices of the caplets that comprise the cap, where each one can be priced by the caplet formula (7.99). A *collar* is a cap and a floor, and can also be priced as such. The concepts used in this section involving selecting the right set of units (or change of numeraire) can even be used to derive convenient closed form solutions for swaptions, which are options on swaps. To learn about the array of interest rate products that can be priced in this manner, the reader is referred to the textbook of Hull [28].

7.6   HEATH–JARROW–MORTON MODEL

Finally, let's work our way to the Heath–Jarrow–Morton (HJM) model [25]. This model can be seen as a limiting case of the LMM, but it was actually developed before the LIBOR Market Model. This time we return to a continuous compounding model of interest rates with the notation $f(t|T_1, T_2)$. Furthermore, we consider discount factors and forward rates over instantaneously small periods of time dt. However, to start we will consider small discrete periods of time Δs, which, upon taking the limit $\Delta s \to 0$, will eventually lead to the instantaneous models.

Since we are starting with discrete periods of time (even if they are small), we are in an identical situation to the LIBOR Market Model. Let's select a fixed maturity time T, and discretize using n time steps $t = s_0 < s_1 < \cdots < s_i < \cdots < s_n = T$ separated by $\Delta s = \frac{T-t}{n}$. Again, at the end we will take limits and get rid of this.

The one difference from the LMM is that we will quote the forward interest rates using a continuous compounding convention. To ease notation, let's follow our convention in the LMM and indicate different rates and discount factors by their final maturity argument. That is, we will denote the forward rates by

$$f(t|s_i) = f(t|s_i - \Delta s, s_i), \tag{7.101}$$

with corresponding discount factors,

$$D_i = D(t|s_i - \Delta s, s_i) = \exp(-f(t|s_i)\Delta s), \quad i = 1 \ldots n. \tag{7.102}$$

In this section I will only consider the question of calibration using the HJM model. In general, the "change-of-units" trick won't reduce common derivatives to a simple form under the HJM model, so we won't pursue that direction. The HJM model is quite general and flexible, but it does not always make pricing the easiest task. In fact, in a lot of ways, the risk neutral pricing approach, which will be introduced in Chapter 10, makes more sense when dealing with the HJM model.

In any case, let's charge ahead with our three step procedure.

Step 1: Model and Classify Variables
The forward rates $f(t|s_i)$ as defined in (7.101) will serve as our underlying variables. Bonds are our tradables, and are related to the forward rates through the equation,

$$B_i = B(t|s_i) = \exp\left(-\sum_{i=1}^{n} f(t|s_i)\Delta s\right) = \prod_{i=1}^{n} D_i,$$

which is the same as equation (7.3) presented at the beginning of this chapter. Furthermore, we will model the forward rates via the stochastic differential equations

$$df(t|s_i) = a(s_i)dt + b(s_i)dz, \tag{7.103}$$

where $a(s_i)$ and $b(s_i)$ can be hiding dependence on time and other variables (i.e., we could have $a(s_i) = a(f(t|s_i), t|s_i)$ and $b(s_i) = b(f(t|s_i), t|s_i)$). Note that I have chosen to use only a single Brownian motion factor dz instead of separate factors dz_i to drive each $f(t|s_i)$. This is similar to assuming $\rho = 1$ between all the factors dz_i in the LMM. Our classification of variables is then

$$
\begin{array}{ccc}
\text{Tradables} & \text{Underlying Variables} & \text{Factors} \\
B_i, i = 1 \ldots n & f(s_i), i = 1 \ldots n & dt, dz.
\end{array}
$$

Step 2 and 3: Use LMM Results

If we think a second before diving headlong into Steps 2 and 3, we will realize that they have already been completed for us in the LMM in terms of discount factors. That is, we know from equation (7.80) that

$$
g_i + h_i \left(\sum_{k=1}^{i-1} \rho_{ik} h_k \right) = h_i \lambda_i, \quad i = 1 \ldots n, \tag{7.104}
$$

where in the HJM case I have chosen to use a single Brownian motion factor. This means that $\rho_{ik} = 1$ and we have only a single λ corresponding to dz. Thus, our calibration equation reduces to

$$
g_i + h_i \left(\sum_{k=1}^{i-1} h_k \right) = h_i \lambda. \tag{7.105}
$$

Now, all we really need to do is follow the lead of Section 7.5.3 and write this calibration equation in terms of the parameters describing $df(t|s_i)$. To do this, we simply apply Ito's lemma to equation (7.102) to give

$$
dD_i = \left(-(\Delta s)a(s_i) + \frac{1}{2}(\Delta s)^2 b^2(s_i) \right) D_i dt - (\Delta s)b(s_i)D_i dz.
$$

Thus, we map the discount factor parameters to those in $df(s_i)$ as

$$
g_i = \left(-(\Delta s)a(s_i) + \frac{1}{2}(\Delta s)^2 b^2(s_i) \right), \quad h_i = -(\Delta s)b(s_i).
$$

Now let's substitute this in equation (7.105) but with $i = n$ so that $s_n = T$ is our fixed maturity time. We have

$$
\left(-(\Delta s)a(s_n) + \frac{1}{2}(\Delta s)^2 b^2(s_n) \right) - (\Delta s)b(s_n) \left(\sum_{i=1}^{n-1} -(\Delta s)b(s_i) \right) = -(\Delta s)b(s_n)\lambda.
$$

Finally, dividing through by Δs, taking the limit as $n \to \infty$ so that $\Delta s \to 0$, and converting the sum to its limiting integral leads to

$$
a(T) - b(T) \int_t^T b(s)ds = b(T)\lambda. \tag{7.106}
$$

Since T was arbitrary, this must hold for any fixed $T > t$. This is the calibration equation for the HJM model. Since I sent $\Delta s \to 0$, the discrete forward rates that spanned a discrete time period Δs turned into instantaneous forward rates in the limit. That is, the HJM is in terms of a continuum of instantaneous forward rates as defined in Section 7.2.5,

$$f(t|s) = \lim_{\Delta s \to 0} f(t|s - \Delta s, s),$$

that follow

$$df(t|s) = a(s)dt + b(s)dz.$$

Here, I short circuited Steps 2 and 3 by using what we had already done for forwards/discount factors in the LMM. However, if you would like to see Steps 2 and 3 done directly without appealing to our forward/discount factor knowledge, then see Appendix 7.9. Venture there at your own peril!

HJM is a general and flexible interest rate model that, quite frankly, is probably best handled in the risk neutral pricing framework. I believe our introduction here is instructive nevertheless.

7.7 CREDIT DERIVATIVES

In this section, we provide a very brief introduction to absence of arbitrage equations for credit derivatives. Credit derivatives usually refer to derivatives that pay off depending on whether a bankruptcy has occurred. In general, there are two main approaches to the modeling of credit derivatives: the structural approach and the reduced form approach. In the structural approach, the actual structure of a company including its debts and other factors, is explicitly taken into account in the modeling. On the other hand, in the reduced form approach, no such detailed modeling is considered. Rather, credit events such as bankruptcies are modeled as sudden Poisson arrivals without providing any explicit economic explanation for their occurrence. In this section, we will limit ourselves to a brief look at some basic reduced form models.

Since a bankruptcy is a sudden event, credit models rely heavily on Poisson processes. In what follows, I will present the simplest model of a defaultable bond when the intensity of default is a constant, and then generalize to allow the intensity to be random. Thus, we will just scratch the surface, but hopefully it will be enough to convey the key ideas. You should note the similarity between these models and Merton's jump diffusion model in Chapter 6, Section 6.6. In fact, if you understand Merton's model then you already have a excellent head start toward understanding these credit based models.

7.7.1 Defaultable Bonds

To tackle models of defaultable bonds, we will build upon a risk-free term structure model. In particular, we will use the single factor short rate model of Section 7.3.2, and then add the possibility of default. As always, we will work through our three step procedure to derive the absence of arbitrage equations. We begin by modeling and classifying the variables.

Step 1: Model and Classify Variables

In deriving equations for defaultable bonds, we need two random factors. The first factor dz drives the short rate process, r_0 as in Section 7.3.2, and the second is a new Poisson default factor $d\pi(\alpha)$ that drives a variable $N(t)$ that indicates default. These are modeled as

$$
\begin{align}
dr_0 &= adt + bdz, \tag{7.107}\\
dN &= d\pi(\alpha). \tag{7.108}
\end{align}
$$

In this case, a defaultable bond of maturity T is a function of r_0, N, and t which we write as $\tilde{B}(r_0, N, t|T)$. Since $N(t)$ is a pure Poisson process, we can start it at $N = 0$ which is the no-default state, and $N = 1$ will be the default state. The idea is that the Poisson process $N(t)$ determines the time of the jump into bankruptcy by when it jumps from $N = 0$ to $N = 1$. By Ito's lemma for Brownian and Poisson we can write, after suppressing all arguments except N,

$$
d\tilde{B}(N) = \tilde{\mathcal{L}}\tilde{B}(N^-)dt + b\tilde{B}_r(N^-)dz + (\tilde{B}(N^- + 1) - \tilde{B}(N^-))d\pi(\alpha)
$$

with

$$
\tilde{\mathcal{L}}\tilde{B}(N^-) = \tilde{B}_t(N^-) + a\tilde{B}_r(N^-) + \frac{1}{2}b^2\tilde{B}_{rr}(N^-).
$$

Since the bond is alive when $N = 0$, and default occurs when $N = 1$, instead of using N^- and $N^- + 1$ in the above equations, let's just substitute $N^- = 0$ and $N^- + 1 = 1$ to give

$$
d\tilde{B}(0) = (\tilde{B}_t(0) + a\tilde{B}_r(0) + \frac{1}{2}b^2\tilde{B}_{rr}(0))dt + b\tilde{B}_r(0)dz + (\tilde{B}(1) - \tilde{B}(0))d\pi(\alpha).
$$

As we did with Merton's jump diffusion model, we will compensate the Poisson jump factor $d\pi(\alpha)$ by subtracting its mean, which is αdt. That is, define a new factor $d\psi = d\pi(\alpha) - \alpha dt$ and rewrite Ito's lemma above as

$$
d\tilde{B}(0) = \mathcal{L}\tilde{B}(0)dt + b\tilde{B}_r(0)dz + (\tilde{B}(1) - \tilde{B}(0))d\psi
$$

with $\mathcal{L}\tilde{B}(0)$ defined as

$$
\mathcal{L}\tilde{B}(0) = \left(\tilde{B}_t(0) + aB_r(0) + \frac{1}{2}b^2B_{rr}(0)\right) + \alpha(B(1) - \tilde{B}(0)).
$$

Our classification of variables is

Tradables	Underlying Variables	Factors
$B_0, \tilde{B}(r_0, N, t\|T)$	$r_0(t), N(t)$	$dt, dz, d\psi.$

Step 2: Construct a Tradable Table

In this case, our tradable table includes the defaultable bond and a money market account. We could also include risk-free zero-coupon bonds, but we will leave them out for now. We have

Prices | Value Change Factor Models

$$
\begin{bmatrix} B_0 \\ \tilde{B}(0) \end{bmatrix} \quad d\begin{bmatrix} B_0 \\ \tilde{B}(0) \end{bmatrix} = \begin{bmatrix} r_0 B_0 \\ \mathcal{L}\tilde{B}(0) \end{bmatrix} dt + \begin{bmatrix} 0 & 0 \\ b\tilde{B}_r(0) & (\tilde{B}(1) - \tilde{B}(0)) \end{bmatrix} \begin{bmatrix} dz \\ d\psi \end{bmatrix}.
$$

where recall again that the argument of $\tilde{B}(N)$ is the default indicator variable N.

Step 3: Apply the Price APT

Finally, we solve the Price APT equation,

$$
\begin{bmatrix} r_0 B_0 \\ \mathcal{L}\tilde{B}(0) \end{bmatrix} = \begin{bmatrix} B_0 \\ \tilde{B}(0) \end{bmatrix} \lambda_0 + \begin{bmatrix} 0 & 0 \\ b\tilde{B}_r(0) & (\tilde{B}(1) - \tilde{B}(0)) \end{bmatrix} \begin{bmatrix} \lambda_1 \\ \lambda_2 \end{bmatrix}.
$$

From the first equation we have $\lambda_0 = r_0$, which is the short rate of interest. The second equation gives

$$
\mathcal{L}\tilde{B}(0) = r_0\tilde{B}(0) + b\tilde{B}_r(0)\lambda_1 + (\tilde{B}(1) - \tilde{B}(0))\lambda_2
$$

which, by replacing $\mathcal{L}\tilde{B}(0)$, can be rewritten as

$$
\tilde{B}_t(0) + a\tilde{B}_r(0) + \tfrac{1}{2}b^2\tilde{B}_{rr}(0) + (\tilde{B}(1) - \tilde{B}(0))\alpha = \\
r_0\tilde{B}(0) + \lambda_1 b\tilde{B}_r(0) + \lambda_2(\tilde{B}(1) - \tilde{B}(0)).
$$

Finally, rearranging leads to

$$
\tilde{B}_t(0) + (a - \lambda_1 b)\tilde{B}_r(0) + \frac{1}{2}b^2\tilde{B}_{rr}(0) = (r_0 + \alpha - \lambda_2)\tilde{B}(0) - (\alpha - \lambda_2)\tilde{B}(1). \quad (7.109)
$$

This is the absence of arbitrage equation for the price of our defaultable bond.

Now, this equation involves two market prices of risk. Where would we determine these market prices of risk? Well, λ_1 is the market price of risk corresponding to the factor driving the *risk-free* term structure of interest rates. Thus, it would be calibrated using risk-free bonds and the single factor short rate models as in Section 7.3.2. The market price of risk λ_2, on the other hand, is associated with the default factor $d\psi = d\pi - \alpha dt$. Thus, to determine its value, it would need to be calibrated to defaultable bonds that depend on the same default risk factor.

7.7.1.1 Different Types of Recovery

Recall that $\tilde{B}(1)$ means that we are in default. Hence, we can assume different recovery values in default by assuming different values for $\tilde{B}(1)$. Examples include no recovery $\tilde{B}(1) = 0$, fractional recovery $\tilde{B}(1) = x\tilde{B}(0)$, where x is the recovery fraction, or even fixed recovery $\tilde{B}(1) = X$ where X is the fixed recovery value upon default.

To see an example, consider the case of no recovery upon default $\tilde{B}(1) = 0$. In this scenario, the absence of arbitrage equation (7.109) becomes

$$\tilde{B}_t(0) + (a - \lambda_1 b)\tilde{B}_r(0) + \frac{1}{2}b^2\tilde{B}_{rr}(0) = (r_0 + \alpha - \lambda_2)\tilde{B}(0). \tag{7.110}$$

The boundary condition if the bond is alive (that is, when $N = 0$) is $\tilde{B}(r_0, N = 0, T|T) = 1$ for a zero-coupon bond.

It turns out that this equation has a simple solution if the corresponding solution for risk-free bonds is known. Let $B(r_0, t|T)$ be the solution in the risk-free no-default setting, then the solution for a defaultable zero-coupon bond is

$$\tilde{B}(r_0, N = 0, t|T) = B(r_0, t|T)e^{-(\alpha-\lambda_2)(T-t)}.$$

Under a continuous compounding convention, this also means that the credit spread (that is, the difference in yield between a risky bond that can default and a risk-free bond) is exactly equal to the intensity of default minus the market price of risk for the Poisson default factor $\alpha - \lambda_2$.

Note that we would expect $\lambda_2 < 0$ since when the Poisson factor jumps up it signals that the bond is defaulting and falling to zero. Thus, this difference in direction between the default factor $d\pi(\alpha)$ and the defaultable bond $\tilde{B}$ means that the market price of risk λ_2 will be negative if there is risk aversion.

Example 7.6 (Term Structure of Defaultable Bonds under Vasicek)
Consider a model of risk-free bonds based on a single factor Vasicek short rate model, as explained in Section 7.3.2.1. Assume that the short rate parameters are $\kappa = 0.25$, $\theta = 0.07$, and $b = 0.02$, and the market price of risk for the Brownian factor driving the short rate process is $\lambda_1 = -0.2$.

Assume that there also exist defaultable bonds with intensity of default $\alpha = 0.02$, Moreover, assume that that market price of risk for the default factor is given by $\lambda_2 = -0.01$.

In this case, if $B(r_0, t|T)$ is the Vasicek price of a risk-free zero-coupon bond, then the price of a defaultable bond is

$$\tilde{B}(r_0, N = 0, t|T) = B(r_0, t|T)e^{-(\alpha-\lambda_2)(T-t)}.$$

For the parameter values given above, the credit spread between the defaultable bonds and risk-free bonds will be exactly $\alpha - \lambda_2 = 0.02 - (-0.01) = 0.03$, or 3%.

A plot of the term structure of interest rates for the risk-free bonds and the defaultable bonds under a continuous compounding convention when $r_0(0) = 5\%$ is given in Figure 7.16. The curves show that the difference in yields is exactly $\alpha - \lambda_2 = 3\%$.

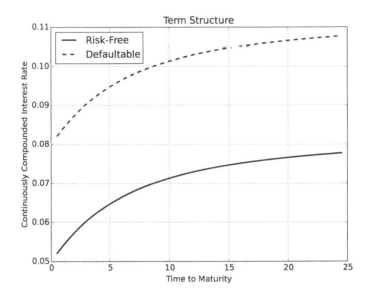

Figure 7.16 Risk-free and defaultable term structure.

7.7.2 Defaultable Bonds with Random Intensity of Default

In the previous section, the intensity of default was a fixed constant α. In this section, we will allow the default intensity to follow a stochastic differential equation of its own and derive the absence of arbitrage equation.

Step 1: Model and Classify Variables

In this setup, we have the following variables:

$$dr_0 \;=\; a\,dt + b\,dz_1, \tag{7.111}$$

$$dN \;=\; d\pi(\alpha), \tag{7.112}$$

$$d\alpha \;=\; f\,dt + g\,dz_2, \tag{7.113}$$

where $\mathbb{E}[dz_1 dz_2] = \rho\,dt$. Again, $r_0(t)$ is the short rate process, $N(t)$ is the Poisson default factor, and $\alpha(t)$ is the default intensity which is now allowed to be random. In this case, a defaultable bond of maturity T is a function of r_0, N, α, and t, denoted $\tilde{B}(r_0, N, \alpha, t | T)$. Again, we start with $N = 0$ which is the no-default state, and let $N = 1$ be the default state. By Ito's lemma (suppressing all arguments except N) we can write

$$d\tilde{B}(0) = \mathcal{L}\tilde{B}(0)dt + b\tilde{B}_r(0)dz_1 + g\tilde{B}_\alpha(0)dz_2 + (\tilde{B}(1) - \tilde{B}(0))d\psi,$$

where the Poisson process has been compensated by defining $d\psi = d\pi(\alpha) - \alpha dt$, and $\mathcal{L}\tilde{B}$ is defined as,

$$\mathcal{L}\tilde{B}(0) = \tilde{B}_t(0) + a\tilde{B}_r(0) + f\tilde{B}_\alpha(0) + \tfrac{1}{2}b^2 \tilde{B}_{rr}(0)$$
$$+ \tfrac{1}{2}g^2 \tilde{B}_{\alpha\alpha}(0) + \rho bg \tilde{B}_{r\alpha}(0) + \alpha(\tilde{B}(1) - \tilde{B}(0)).$$

Our classification of variables is

$$
\begin{array}{ccc}
\text{Tradables} & \text{Underlying Variables} & \text{Factors} \\
B_0, \tilde{B}(N) & r_0(t), \alpha(t), N(t) & dt, dz_1, dz_2, d\psi.
\end{array}
$$

Step 2: Construct a Tradable Table
We create a tradable table with the money market account and the defaultable bonds as follows:

Prices   Value Change Factor Models

$$
\begin{bmatrix} B_0 \\ \tilde{B}(0) \end{bmatrix} \quad
d \begin{bmatrix} B_0 \\ \tilde{B}(0) \end{bmatrix} =
\begin{bmatrix} r_0 B_0 \\ \mathcal{L}\tilde{B}(0) \end{bmatrix} dt +
\begin{bmatrix} 0 & 0 & 0 \\ b\tilde{B}_r(0) & g\tilde{B}_\alpha(0) & (\tilde{B}(1) - \tilde{B}(0)) \end{bmatrix}
\begin{bmatrix} dz_1 \\ dz_2 \\ d\psi \end{bmatrix}.
$$

Step 3: Apply the Price APT
Finally, the Price APT is given by

$$
\begin{bmatrix} r_0 B_0 \\ \mathcal{L}\tilde{B}(0) \end{bmatrix} =
\begin{bmatrix} B_0 \\ \tilde{B}(0) \end{bmatrix} \lambda_0 +
\begin{bmatrix} 0 & 0 & 0 \\ b\tilde{B}_r(0) & g\tilde{B}_\alpha(0) & (\tilde{B}(1) - \tilde{B}(0)) \end{bmatrix}
\begin{bmatrix} \lambda_1 \\ \lambda_2 \\ \lambda_3 \end{bmatrix}.
$$

From the first equation $\lambda_0 = r_0$, which is the short rate. The second equation gives

$$
\mathcal{L}\tilde{B}(0) = r_0 \tilde{B}(0) + b\tilde{B}_r(0)\lambda_1 + g\tilde{B}_\alpha(0)\lambda_2 + (\tilde{B}(1) - \tilde{B}(0))\lambda_3
$$

which, after replacing $\mathcal{L}\tilde{B}(0)$, can be rewritten as

$$
\tilde{B}_t(0) + (a - \lambda_1 b)\tilde{B}_r(0) + (f - \lambda_2 g)\tilde{B}_\alpha(0) + \tfrac{1}{2}b^2 \tilde{B}_{rr}(0) + \tfrac{1}{2}g^2 \tilde{B}_{\alpha\alpha}(0) + \rho g b \tilde{B}_{r\alpha}(0)
$$
$$
= (r_0 + \alpha - \lambda_3)\tilde{B}(0) + (\lambda_3 - \alpha)\tilde{B}(1).
$$

This is the absence of arbitrage equation that governs a defaultable bond with random intensity of default. As explained in the previous section, λ_1 would be calibrated from the risk-free term structure. The other market prices of risk, λ_2 and λ_3, would need to be calibrated from defaultable bonds or securities that depend on the same factors, dz_2 and $d\psi$.

As in the case of a fixed intensity of default, we can consider different possibilities for recovery, such as no recovery $\tilde{B}(1) = 0$, fractional recovery $\tilde{B}(1) = x\tilde{B}(0)$, or fixed recovery $\tilde{B}(1) = X$.

This partial differential equation for absence of arbitrage is more difficult that the corresponding equation under constant intensity of default. While there are certain modeling choices that can lead to convenient solutions for the prices of defaultable zero-coupon bonds, I will not pursue those equations here. Rather, the interested reader is referred to [17, 47].

7.8 SUMMARY

In this chapter, we used the factor approach to derive absence of arbitrage pricing equations for interest rate and credit models. Once again, it was merely an application of our three step procedure, and the purpose was to demonstrate that the approach is generic enough to capture a wide array of the absence of arbitrage equations and relationships that arise.

In particular, we saw that there are two main approaches to modeling the risk-free term structure of interest rates. They are distinguished by whether the short rate process or forward rates are used as the underlying variables. For short rate based models, we presented the Vasicek and CIR models in which closed form solutions are available for zero-coupon bond prices. In addition, many other such single-factor short rate models have been developed.

To explain models of the term structure based on forward rates as underlying variables, we began with a detour into options on forward contracts. The results that we derived for term structure models, including the LIBOR Market Model and the Heath–Jarrow–Morton model, were presented as an extension of the concepts for forward contracts, but applied to forward rates. Finally, we briefly touched upon reduced form credit risk models, where Poisson factors were used to signal the default of a bond.

7.9 APPENDIX: HEATH–JARROW–MORTON DERIVATION

In the Heath–Jarrow–Morton [25] framework, the instantaneous forward rates are taken as the underlying variables. In this appendix, the full three step route is taken to derive the calibration equation for the HJM.

Step 1: Model and Classify Variables

Let $f(t|s)$ denote the backward-differenced instantaneous forward rate under continuous compounding, defined as $f(t|s) = \lim_{\Delta s \to 0} f(t|s - \Delta s, s)$. We use this backward difference convention to be consistent with Section 7.6. In that case, it was an artifact of our notation in the LIBOR Market Model.

Further, recall that the instantaneous short rate $r_0(t)$ can be defined relative to forward rates as $r_0(t) = \lim_{\Delta t \to 0} f(t|t, t + \Delta t)$. Using this connection, we will also use the notation $f(t|t) = r_0(t)$ to indicate the instantaneous short rate.

A zero-coupon bond at time t with maturity T will be denoted by $B(t|T)$ and is related to the instantaneous forward rates by

$$B(t|T) = \exp\left(-\int_t^T f(t|s)ds\right). \tag{7.114}$$

This equation follows directly from the definition of a forward rate and was given previously in equation (7.9). Taking $\ln(\cdot)$ of both sides of this equation and differentiating with respect to T leads to

$$f(t|T) = -\frac{\partial}{\partial T}\ln(B(t|T)). \tag{7.115}$$

HJM takes the instantaneous forward rates as the underlying variables, and models them as

$$df(t|s) = a(s)dt + b(s)dz, \qquad (7.116)$$

where $a(s)$ and $b(s)$ can be hiding dependence on time and other variables.

Note that equation (7.114) gives us an *explicit* relationship between the tradables and the (infinite number of) underlying variables. Thus, instead of using a generic Ito's lemma relationship between $B(t|T)$ and the underlying factors $f(t|s)$, we will be able to make that relationship concrete by plugging explicit partial derivatives into Ito's lemma.

Why is this helpful? Because this will allow us to pull the Price APT relationship for the marketed tradables $B(t|T)$ back to the underlying variables $f(t|s)$ which opens up the possibility of calibration directly on the underlying variables $f(t|s)$ instead of the marketed tradables. If it is easier to work with market data about instantaneous forward rates, then this can simplify the calibration process. To understand the value of this, you should consider the calibration procedure that has to be done for a single or multifactor short rate model if no closed form bond pricing formula is known (i.e., no closed form solution to the pricing partial differential equation is known), and compare that with using forward rate data to calibrate in the HJM model that we will derive.

With those preliminaries out of the way, let's dive into the details. Our classification of variables is

Tradables	Underlying Variables	Factors
$B_0, B(t|T)$	$f(t|s), s \in [t, T]$	dt, dz.

In this setup, we will take zero-coupon bonds as our marketed tradables. That is, $B(t|T)$ is tradable, which means that we need to calculate $dB(t|T)$ for our tradable table. This is a tricky calculation because $B(t|T)$ is really a function of an infinite number of Ito processes $f(t|s)$ for $s \in [t, T]$ through the equation

$$B(t|T) = \exp\left(-\int_t^T f(t|s)ds\right). \qquad (7.117)$$

To simplify our thinking, let's begin by assuming that s is indexed by $k = 1...n$ so that $B(t|T)$ will depend only on n Ito processes. That is, let us assume

$$B(t|T) = B(\{f(t|s_k) : k = 1...n\}, t|T) \qquad (7.118)$$

with $t = s_0 < s_1 < s_2 < \cdots < s_n = T$ and $\Delta s_k = s_k - s_{k-1}$. This allows us to work with a discrete approximation when needed,

$$B(t|T) \approx \exp\left(-\sum_{k=1}^n f(t|s_k)\Delta s_k\right). \qquad (7.119)$$

For notational simplicity, let $B = B(t|T)$. Then, by Ito's lemma applied to (7.119), we have

$$dB = \left(B_t + \sum_{i=1}^{n} a(s_i)B_{f(t|s_i)} + \sum_{i=1}^{n}\sum_{j=1}^{n} \frac{1}{2}b(s_i)b(s_j)B_{f(t|s_i)f(t|s_j)} \right) dt \qquad (7.120)$$
$$+ \left(\sum_{i=1}^{n} b(s_i)B_{f(t|s_i)} \right) dz.$$

To complete our computation of Ito's lemma, we need to compute the partial derivatives B_t, $B_{f(t|s_i)}$, and $B_{f(t|s_i)f(t|s_j)}$. To compute B_t we can use the continuous model in equation (7.117) directly to obtain

$$B_t = \exp\left(-\int_t^T f(t|s)ds \right) f(t|t) = B(t|T)r_0(t), \qquad (7.121)$$

where the second equality used equation (7.117) and changed notation for the instantaneous short rate from $f(t|t)$ to $r_0(t)$. (Note that no differentiation under the integral is necessary here because we are considering the partial derivative with respect to t. The variables $f(t|s)$ are considered separate for the purpose of the partial derivative. Thus, even though we see a t in them, that does not play into the partial derivative.)

For $B_{f(t|s_i)}$ and $B_{f(t|s_i)f(t|s_j)}$ we use the discretization in (7.119), which leads to

$$B_{f(t|s_i)} \approx -\exp\left(-\sum_{k=1}^{n} f(t|s_k)\Delta s_k \right) \Delta s_i$$
$$B_{f(t|s_i)f(t|s_j)} \approx \exp\left(-\sum_{k=1}^{n} f(t|s_k)\Delta s_k \right) \Delta s_i\Delta s_j.$$

Substituting these into Ito's lemma in (7.120) gives

$$dB(t|T) = \left(B(t|T)r_0(t) - \sum_{i=1}^{n} a(s_i)\exp\left(-\sum_{k=1}^{n} f(t|s_k)\Delta s_k \right) \Delta s_i \right.$$
$$\left. + \frac{1}{2}\sum_{i=1}^{n}\sum_{j=1}^{n} b(s_i)b(s_j)\exp\left(-\sum_{k=1}^{n} f(t|s_k)\Delta s_k \right) \Delta s_i\Delta s_j \right) dt$$
$$- \left(\sum_{i=1}^{n} b(s_i)\exp\left(-\sum_{k=1}^{n} f(t|s_k)\Delta s_k \right) \Delta s_i \right) dz.$$

Taking continuous limits yields

$$
\begin{aligned}
dB(t|T) &= \left(B(t|T)r_0(t) - B(t|T)\int_t^T a(s)ds \right. \\
&\qquad \left. +\frac{1}{2}B(t|T)\int_t^T\int_t^T b(s)b(r)drds \right)dt - \left(B(t|T)\int_t^T b(s)ds \right)dz \\
&= B(t|T)\left[\left(r_0(t) - \int_t^T a(s)ds + \frac{1}{2}\int_t^T\int_t^T b(s)b(r)drds \right)dt \right. \\
&\qquad \left. - \left(\int_t^T b(s)ds \right)dz \right] \\
&= B(t|T)\left[\tilde{\mu}(T)dt - \left(\int_t^T b(s)ds \right)dz \right] \qquad (7.122)
\end{aligned}
$$

where

$$
\tilde{\mu}(T) = \left(r_0(t) - \int_t^T a(s)ds + \frac{1}{2}\int_t^T\int_t^T b(s)b(r)drds \right). \qquad (7.123)
$$

Equation (7.122) describes our tradable bonds.

Step 2: Construct a Tradable Table
The above modeling allows us to construct our tradable table as

$$
\begin{array}{c|c}
\text{Prices} & \text{Value Change Factor Models} \\
\hline
\begin{bmatrix} B_0 \\ B(t|T) \end{bmatrix} & d\begin{bmatrix} B_0 \\ B(t|T) \end{bmatrix} = \begin{bmatrix} r_0(t)B_0 \\ \tilde{\mu}(T)B(t|T) \end{bmatrix}dt + \begin{bmatrix} 0 \\ -B(t|T)\left(\int_t^T b(s)ds\right) \end{bmatrix}dz,
\end{array}
$$

where B_0 is the money market account.

Step 3: Apply the Price APT
Applying the Price APT equation leads to

$$
\begin{bmatrix} r_0(t)B_0 \\ \tilde{\mu}(T)B(t|T) \end{bmatrix} = \begin{bmatrix} B_0 \\ B(t|T) \end{bmatrix}\lambda_0 + \begin{bmatrix} 0 \\ -B(t|T)\left(\int_t^T b(s)ds\right) \end{bmatrix}\lambda.
$$

Solving the first equation gives $\lambda_0 = r_0(t)$, while the second equation leads to

$$
r_0(t) - \int_t^T a(s)ds + \frac{1}{2}\int_t^T\int_t^T b(s)b(r)drds = r_0(t) - \left(\int_t^T b(s)ds \right)\lambda
$$

or

$$
\int_t^T a(s)ds - \frac{1}{2}\int_t^T\int_t^T b(s)b(r)drds = \left(\int_t^T b(s)ds \right)\lambda.
$$

This equation looks slightly different from equation (7.106). However, to match them up we just need to take the partial derivative with respect to T, giving

$$a(T) - b(T) \int_t^T b(s)ds = b(T)\lambda. \tag{7.124}$$

The upshot of this calculation is that the expected drift rate and standard deviation of the instantaneous forward rates $f(t|s)$ must be related through this calibration equation.

EXERCISES

7.1 Complete the table below by converting the term structure of spot rates under continuous compounding to zero-coupon bond prices, forward rates, and discount factors.

Time: T_i	0.1	0.2	0.3	0.4
Spot Rate: $r(0|T_i)$	3.1%	3.2%	3.25%	3.28%
Zero-Coupon Bonds: $B(0|T_i)$	?	?	?	?
Forward Rate: $f(0|T_{i-1}|T_i)$	?	?	?	?
Discount Factors: $D(0|T_{i-1}, T_i)$	?	?	?	?

7.2 Complete the table below by converting the zero-coupon bond prices to spot rates, forward rates, and discount factors.

Time: T_i	0.1	0.2	0.3	0.4
Spot Rate: $r(0|T_i)$	?	?	?	?
Zero-Coupon Bonds: $B(0|T_i)$	0.99	0.98	0.97	0.96
Forward Rate: $f(0|T_{i-1}|T_i)$	?	?	?	?
Discount Factors: $D(0|T_{i-1}, T_i)$	?	?	?	?

7.3 Complete the table below by converting the forward rates to discount factors, bond prices, and spot rates.

Time: T_i	0.2	0.4	0.6	0.8
Spot Rate: $r(0|T_i)$	?	?	?	?
Zero-Coupon Bonds: $B(0|T_i)$	?	?	?	?
Forward Rate: $f(0|T_{i-1}|T_i)$	5.00%	5.04%	5.06%	5.07%
Discount Factors: $D(0|T_{i-1}, T_i)$	?	?	?	?

7.4 Complete the table below by converting the discount factors to forward rates, bond prices, and spot rates.

Time: T_i	0.2	0.4	0.6	0.8
Spot Rate: $r(0|T_i)$	?	?	?	?
Zero-Coupon Bonds: $B(0|T_i)$	?	?	?	?
Forward Rate: $f(0|T_{i-1}|T_i)$	?	?	?	?
Discount Factors: $D(0|T_{i-1}, T_i)$	0.99	0.98	0.98	0.97

7.5 Compute the price of a zero-coupon bond at time $t = 0$ using the Vasicek model with the following parameter values:

(a) $r_0 = 0.04$, $\kappa = 0.3$, $\theta = 0.06$, $b = 0.03$, $T = 0.5$ and $\lambda = -0.1$.

(b) $r_0 = 0.06$, $\kappa = 0.5$, $\theta = 0.03$, $b = 0.01$, $T = 0.25$ and $\lambda = -0.3$.

(c) $r_0 = 0.01$, $\kappa = 0.4$, $\theta = 0.08$, $b = 0.02$, $T = 2$ and $\lambda = -0.2$.

(d) $r_0 = 0.1$, $\kappa = 0.6$, $\theta = 0.08$, $b = 0.04$, $T = 1.5$ and $\lambda = -0.2$.

7.6 Forward-Spot Relationship.

Compute the absence of arbitrage spot price of an asset, $S(0)$, given the forward price $F(0|T)$ and spot rate of interest under continuous compounding $r(0|T)$. Use the following values:

(a) $T = 1$, $F(0|1) = \$10$ and $r(0|1) = 0.05$.

(b) $T = 5$, $F(0|5) = \$25$ and $r(0|5) = 0.04$.

(c) $T = 0.5$, $F(0|0.5) = \$18$ and $r(0|0.5) = 0.03$.

(d) $T = 0.75$, $F(0|5) = \$100$ and $r(0|5) = 0.08$.

7.7 Option on a Forward.

Consider the forward price, $f(t|T)$, of an asset that follows

$$df = \mu_f f dt + \sigma_f f dz,$$

and let $r(t|T)$ be the continuously compounded spot rate for time T. Use Black's Model to compute the value of a European call option at time $t = 0$ on a forward under the following parameter values,

(a) $T = 0.5$, $K = \$10$, $f(0|T) = \$9.5$, $\sigma_f = 0.25$, and $r(0|T) = 0.04$.

(b) $T = 0.2$, $K = \$51$, $f(0|T) = \$50$, $\sigma_f = 0.3$, and $r(0|T) = 0.07$.

(c) $T = 0.3$, $K = \$24$, $f(0|T) = \$25$, $\sigma_f = 0.4$, and $r(0|T) = 0.03$.

(d) $T = 0.75$, $K = \$100$, $f(0|T) = \$100$, $\sigma_f = 0.25$, and $r(0|T) = 0.06$.

7.8 LMM Caplet Formula.

Assume that all loans have a principal of $\$1,000,000$. The interest rate is set at time T_{i-1} and the interest payment is made at time $T_i = T_{i-1} + \tau$. Use the LMM caplet formula to price caplets at time $t = 0$ under the following parameter values,

(a) $T_{i-1} = 0.75$, $\tau = 0.25$, $F(0|T_{i-1}, T_i) = 4\%$, $b_{T_i} = 20\%$, $K = 3.8\%$, $B(0|T_i) = 0.97409$.

(b) $T_{i-1} = 0.6$, $\tau = 0.2$, $F(0|T_{i-1}, T_i) = 5\%$, $b_{T_i} = 15\%$, $K = 5.10\%$, $B(0|T_i) = 0.96233$.

(c) $T_{i-1} = 1.25$, $\tau = 0.25$, $F(0|T_{i-1}, T_i) = 8\%$, $b_{T_i} = 25\%$, $K = 8\%$, $B(0|T_i) = 0.88825$.

(d) $T_{i-1} = 1.5$, $\tau = 0.5$, $F(0|T_{i-1}, T_i) = 6\%$, $b_{T_i} = 15\%$, $K = 5.75\%$, $B(0|T_i) = 0.88870$.

7.9 Give a formula for the spot rate under continuous compounding in the Vasicek model. Use the formula to compute the spot rate under Vasicek dynamics for the parameter values, $t = 0$, $r_0(0) = 0.04$, $\kappa = 0.3$, $\theta = 0.06$, $b = 0.03$, $T = 0.5$ and $\lambda = -0.1$.

7.10 Derive the absence of arbitrage equation for the price of a European call option on a non-dividend paying stock when interest rates are random, and the short rate follows Cox–Ingersoll–Ross dynamics:

$$dS = \mu S dt + \sigma S dz_1 \qquad (P7.1)$$
$$dr_0 = \kappa(\theta - r_0)dt + b\sqrt{r_0}dz_2 \qquad (P7.2)$$

where dz_1 and dz_2 are correlated $\mathbb{E}[dz_1 dz_2] = \rho dt$. Your answer may contain the market price of risk for dz_2.

7.11 Flat Term Structure.

Consider a model in which the term structure of interest rates is flat, but moves up and down randomly. That is, let zero-coupon bond prices with face value \$1 and maturity T be denoted by $B(t|T)$ and satisfy

$$B(t|T) = \exp(-r(T - t)) \qquad (P7.3)$$

where r is modeled as a stochastic differential equation,

$$dr = a(r, t)dt + b(r, t)dz. \qquad (P7.4)$$

Note that this model applies for all maturities T.

(a) Write models for the instantaneous return of the money market account and for a generic zero-coupon bond with maturity T (i.e., for $B(t|T)$).

(b) Derive the absence of arbitrage condition in this case and derive restrictions on $a(r, t)$ and $b(r, t)$. What does this imply about the allowable dynamics for this term structure? In particular, argue that the only possible flat absence of arbitrage term structure is a fixed constant.

7.12 Defaultable Bond Solution.

Consider equation (7.110) for a defaultable bond when the term structure of risk-free bonds is given by a single factor short rate model:

$$\tilde{B}_t(0) + (a - \lambda_1 b)\tilde{B}_r(0) + \frac{1}{2}b^2 \tilde{B}_{rr}(0) = (r_0 + \alpha - \lambda_2)\tilde{B}(0)$$

and there is no recovery at default.

(a) Let $B(r_0, t|T)$ be the price of a risk-free bond with maturity T. The boundary condition is $B(r, T|T) = \$1$. Show that $\tilde{B}(0) = \tilde{B}(r_0, N = 0, t|T) = B(r_0, t|T)e^{-(\alpha-\lambda_2)(T-t)}$ is the price of a defaultable bond with payoff of $\$1$ at maturity and no recovery after default. That is, show that it solves the equation above with boundary condition $\tilde{B}(r_0, N = 0, T|T) = 1$ if $B(r_0, t|T)$ solves the corresponding equation

$$B_t(T) + (a - \lambda_1 b)B_r(T) + \frac{1}{2}b^2 B_{rr}(T) = r_0 B(T)$$

for a risk-free bond.

(b) Let the short rate process be given by the Vasicek model with parameter values, $t = 0$, $a = \kappa(\theta - r_0)$, $\kappa = 0.3$, $\theta = 0.06$, $b = 0.03$, $r_0(0) = 0.04$, $T = 0.5$ and $\lambda_1 = -0.1$. Assume that the intensity of default is $\alpha = 0.02$ and the market price of the default risk is $\lambda_2 = -0.01$. What is the price of this defaultable bond?

7.13 Options on Forwards and Futures.

Consider the following dynamics,

$$\begin{aligned} df &= \mu f dt + \sigma f dz_1, \\ dr_0 &= a dt + b dz_2, \end{aligned}$$

where f is either the forward price or futures price and r_0 is the random instantaneous short rate. The correlation between dz_1 and dz_2 is assumed to be $\mathbb{E}[dz_1 dz_2] = \rho dt$.

(a) Derive the absence of arbitrage equation for a derivative on a futures contract. Let the derivative take the functional form $c(f, r_0, t)$.

(b) Derive the absence of arbitrage equation for a derivative on a forward contract. Let the derivative take the functional form $c(f, r_0, t)$. (Hint: This was already essentially done in Section 7.4.1.2 just with different notation. Remember to use the spot price of the asset as the tradable and assume that the bond is driven by dz_2.)

(c) What if interest rates are not stochastic (that is, if $b = 0$)? Do the absence of arbitrage equations for options on forwards and futures look the same? What if the correlation satisfies $\rho = 0$? Again, do the absence of arbitrage equations for options on forwards and futures look the same?

CHAPTER 8

Hedging

To hedge means to reduce risk. In the context of derivative pricing, the goal of hedging is to combine the derivative security with other assets in order to minimize the risk of the resulting portfolio.

In this chapter, we present two main approaches to hedging. In the first, we recognize that risk comes from the factors. Thus, hedging is based on eliminating factor risk. In the second approach we consider a derivative to be a function of underlying variables. We are then hedged against variations in these underlying variables if the (calculus-based) derivative of the value of our portfolio with respect to the underlying variables is zero. That is, our portfolio is not sensitive to small variations in the underlying variables.

8.1   HEDGING FROM A FACTOR PERSPECTIVE

In abstract terms, hedging involves eliminating factor risk. Returning to our basic modeling paradigm and classification of variables, in this section we view random factors as the underlying drivers of risk. Thus, if we can eliminate a factor from a portfolio, then we have hedged against that specific risk. This perspective on hedging is diagrammed in Figure 8.1.

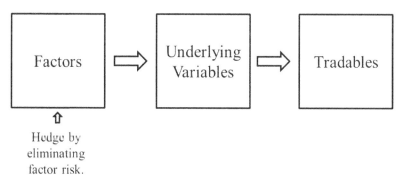

Figure 8.1   Hedging from a factor perspective.

Hedging problems usually begin by identifying what you would like to hedge, and what you can hedge with. In our context, we would like to hedge the risk in a derivative security. More specifically, let dV_d be the value change per unit of a derivative security which follows the factor model

$$dV_d = A_d dt + B_d dz. \tag{8.1}$$

We assume that we would like to hedge the risk (which comes from the factors dz) in a *single* unit of this derivative.

Next, we need to identify what we can hedge with. Here, our assumption is that we can hedge with marketed tradables whose value changes are governed by

$$dV_m = A_m dt + B_m dz. \tag{8.2}$$

(Recall that dV_m can be a vector if there is more than one marketed tradable.)

To create an actual hedge using the marketed tradables we need to specify how much of each marketed tradable we will hold. In this regard, let y_m be a vector whose elements indicate the number of units of each marketed tradable. Our hedged portfolio will then contain a single unit of the derivative combined with y_m units of the marketed tradables. The value change of our overall hedged portfolio dV_h (the subscript h will be used to denote *hedged* quantities) will simply be the combination of the value change of the derivative dV_d plus the value change caused by our holdings in the marketed tradables $y_m^T dV_m$:

$$dV_h = dV_d + y_m^T dV_m = (A_d + y_m^T A_m)dt + (B_d + y_m^T B_m)dz. \tag{8.3}$$

Since the goal of hedging is to eliminate risk, and risk comes from the factors dz, we can simply zero-out the coefficients of the factors dz in (8.3) to eliminate their effect on the hedged portfolio. That is, we should choose y_m so that

$$B_d + y_m^T B_m = 0. \tag{8.4}$$

This is really all there is to hedging!

8.1.1 Description Using a Tradable Table

I like to use tradable table descriptions when possible, so let's rework the hedging analysis in the structure of a tradable table. Consider a standard tradable table description, but augment it with our holdings variable y. That is, y is a vector in which each component contains the number of units of each asset held in our portfolio. It will look like this:

$$
\begin{array}{ccc}
\text{Holdings} & \text{Prices} & \text{Value Change Factor Models} \\
--- | --- & | & ---------------\\
[\, y \,] & [\, \mathcal{P} \,] & d[\, \mathcal{V} \,] = [\, \mathcal{A} \,]dt + [\, \mathcal{B} \,]dz.
\end{array}
$$

Using our standard convention, assume that the last tradable is the derivative that we would like to hedge. Moreover, assume that we hold one unit of that derivative. Thus, our augmented tradable table takes the form,

$$
\begin{array}{ccc}
\text{Holdings} & \text{Prices} & \text{Value Change Factor Models} \\
--- \,| & ----\,| & --------------- \\
\begin{bmatrix} y_m \\ -- \\ 1 \end{bmatrix} &
\begin{bmatrix} \mathcal{P}_m \\ -- \\ \mathcal{P}_d \end{bmatrix} &
d\begin{bmatrix} \mathcal{V}_m \\ -- \\ \mathcal{V}_d \end{bmatrix} =
\begin{bmatrix} \mathcal{A}_m \\ -- \\ \mathcal{A}_d \end{bmatrix} dt +
\begin{bmatrix} \mathcal{B}_m \\ -- \\ \mathcal{B}_d \end{bmatrix} dz.
\end{array}
$$

In this case, as can be seen in the tradable table, our total holdings vector actually contains the structure

$$
y = \begin{bmatrix} y_m \\ 1 \end{bmatrix}
\tag{8.5}
$$

and the value change in the hedged portfolio is simply computed as

$$
d\mathcal{V}_h = y^T d\mathcal{V} = y^T \mathcal{A} dt + y^T \mathcal{B} dz.
\tag{8.6}
$$

Thus, to hedge, we select y so that $y^T \mathcal{B} = 0$ (recalling of course that y is constrained to the structure given by (8.5)). This is exactly the analysis given in the previous section.

8.1.2 Relationship between Hedging and Arbitrage

Hedging is actually quite closely related to arbitrage. Recall the arbitrage price implication from Chapter 4, Section 4.3.2:

$$
\left.\begin{array}{ll}
y^T \mathcal{P} = 0 & \text{No cost} \\
y^T \mathcal{B} = 0 & \text{No risk}
\end{array}\right\} \Rightarrow y^T \mathcal{A} = 0 \quad \text{No profit/loss.}
\tag{8.7}
$$

Note that $y^T \mathcal{B} = 0$ is one of the conditions of the arbitrage price implication, which turns out to be identical to what we desire from a hedged portfolio. Therefore, a hedged portfolio will automatically satisfy this condition.

The other condition that needs to be satisfied to set up an arbitrage is that the total cost or price must be zero, $y^T \mathcal{P} = 0$. Let's see how we can alter our hedged portfolio to satisfy this no cost condition.

8.1.2.1 Creating a No-Cost Hedge

Let's break out the tradable table in even more detail. First, assume that there is a tradable with no direct factor risk:

$$
dB_0 = r_0 B_0 dt.
\tag{8.8}
$$

This is our risk-free bond (or money market account depending on the situation), and I will explicitly separate it from the other marketed tradables. The other marketed tradables (excluding the risk-free bond) used to hedge this asset will be denoted by

$$d\hat{\mathcal{V}} = \hat{\mathcal{A}}dt + \hat{\mathcal{B}}dz. \tag{8.9}$$

As before, the derivative to be hedged satisfies

$$d\mathcal{V}_d = \mathcal{A}_d dt + \mathcal{B}_d dz. \tag{8.10}$$

We can compile this information in an augmented tradable table as

$$
\begin{array}{ccc}
\text{Holdings} & \text{Prices} & \text{Value Change Factor Models} \\
--- | ---- & | & ----------------
\end{array}
$$

$$
\begin{bmatrix} y_0 \\ \hat{y} \\ -- \\ 1 \end{bmatrix}
\quad
\begin{bmatrix} B_0 \\ \hat{\mathcal{P}} \\ -- \\ \mathcal{P}_d \end{bmatrix}
\quad
d\begin{bmatrix} B_0 \\ \hat{\mathcal{V}} \\ -- \\ \mathcal{V}_d \end{bmatrix}
=
\begin{bmatrix} r_0 B_0 \\ \hat{\mathcal{A}} \\ -- \\ \mathcal{A}_d \end{bmatrix} dt
+
\begin{bmatrix} 0 \\ \hat{\mathcal{B}} \\ -- \\ \mathcal{B}_d \end{bmatrix} dz,
$$

where y_0 is the holding of B_0 and $\hat{y}$ represents the holdings of the other marketed tradables. Thus, our total holdings in the the marketed tradables has been broken into $y_m = [y_0, \hat{y}^T]^T$.

Now, we make the following important observation. The holding in the first asset y_0 has no effect on the hedge! Since it has no factor risk, including it in a portfolio has no effect on the factors. Hence, we can choose its holding y_0 arbitrarily without affecting the hedge.

This recognition allows us to select y_0 so that the hedge has zero cost. That is, let $\hat{y}$ be the holdings of a hedged portfolio and select y_0 so that the total cost is zero:

$$y_0 B_0 + \hat{y}^T \hat{\mathcal{P}} + \mathcal{P}_d = 0. \tag{8.11}$$

Thus, as long as a risk-free bond or money market account exists with no direct factor risk, any completely hedged portfolio can be altered to satisfy the left side conditions of the Price APT implication (8.7). Thus, for no arbitrage to exist, we must have zero profit/loss in this hedged portfolio. This means

$$y_0 r_0 B_0 + \hat{y}^T \hat{\mathcal{A}} + \mathcal{A}_d = 0. \tag{8.12}$$

This profit/loss condition is actually our absence of arbitrage equation! For example, a hedging strategy can be used to derive the Black–Scholes equation via the above relationships.

8.1.2.2 Simple Explanation

Here is the simple explanation of what we have done above. We hold an asset that we would like to hedge. First, we use $\hat{y}$ to hedge away all the factor risk in our portfolio. Thus, the hedged portfolio has no direct factor risk. Since it has no factor risk, for no arbitrage to exist it must be the same as the first tradable that also has no factor risk. In the Black–Scholes setting, this first asset is the risk-free asset, and all we are saying is that the hedged portfolio must earn the risk-free rate. In an interest rate derivative setting, the tradable without direct factor risk is the money market account and we are saying that the hedged portfolio must earn the short rate. That is the essence of the above calculations!

8.1.3 Hedging Examples

We have presented all this in abstract terms, so let's see how hedging is done in some specific examples.

8.1.3.1 Hedging in Black–Scholes

In the Black–Scholes setup of Chapter 6, Section 6.1 we have the following augmented tradable table with the first column showing the holdings:

Holdings   Prices   Value Change Factor Models

$$
\begin{bmatrix} y_0 \\ \hat{y} \\ - \\ 1 \end{bmatrix} \quad \begin{bmatrix} B_0 \\ S \\ - \\ c \end{bmatrix} \quad d \begin{bmatrix} B_0 \\ S \\ - \\ c \end{bmatrix} = \begin{bmatrix} r_0 B_0 \\ \mu S \\ - - - \\ c_t + \mu S c_S + \frac{1}{2}\sigma^2 S^2 c_{SS} \end{bmatrix} dt + \begin{bmatrix} 0 \\ \sigma S \\ - - \\ \sigma S c_S \end{bmatrix} dz.
$$

Note that we have assumed that there is one option c and that we are hedging with the stock S. Thus, we assume the holding (shares) of the stock is $\hat{y}$. This portfolio is hedged if $y^T \mathcal{B} = 0$, which in this case reduces to

$$
\hat{y}\sigma S + \sigma S c_S = 0. \tag{8.13}
$$

Solving for $\hat{y}$ gives $\hat{y} = -c_S$, which must be evaluated at the current value of S. Thus, we have a hedged portfolio if we hold $-c_S$ shares of the stock for every option. The quantity c_S is typically called the *delta* of the option.

8.1.3.2 Pricing from the Black–Scholes Hedge

The above portfolio is hedged, but we can convert it to satisfy the arbitrage implication by choosing y_0 so that the total cost is zero:

$$
y_0 B_0 - c_S S + c = 0. \tag{8.14}
$$

This is now a potential arbitrage portfolio. For no arbitrage to be present, we must have no profit/loss which means that the drift term of our portfolio must be zero:

$$
y_0 r_0 B_0 - c_s \mu S + (c_t + \mu S c_S + \frac{1}{2}\sigma^2 S^2 c_{SS}) = 0. \tag{8.15}
$$

Substituting in (8.15) for $y_0 B_0 = c_S S - c$ from (8.14) gives

$$
r_0 (c_S S - c) - c_s \mu S + (c_t + \mu S c_S + \frac{1}{2}\sigma^2 S^2 c_{SS}) = 0 \tag{8.16}
$$

or

$$c_t + r_0 S c_S + \frac{1}{2}\sigma^2 S^2 c_{SS} = r_0 c, \qquad (8.17)$$

which is the Black–Scholes equation.

Once again, the simple explanation of this derivation is that the hedged portfolio is risk free and hence must earn the risk-free rate. I presented the more structured derivation because I believe that it is always important to have structure to fall back on when your intuition fails. However, the intuition is important as well!

Example 8.1 (Delta for European Calls and Puts)
Consider European call and put options with strike $K = \$10$ and time to maturity $T = 0.3$ on a non-dividend paying stock S. Assume the stock follows geometric Brownian motion with volatility $\sigma = 0.3$, and the risk-free rate is $r_0 = 0.05$.

Referring to the Black–Scholes formulas for calls and puts in equations (6.7) and (6.10) of Chapter 6, Section 6.1, we can compute the amount to hold in the underlying stock, S, to hedge the call or put as

$$\hat{y} = -c_S = -\Phi(d_1)$$

for the call option, and

$$\hat{y} = -p_S = \Phi(-d_1)$$

for the put. Note that the hedge quantity $\hat{y}$ is minus the partial derivative of the value of an option with respect to the stock price. This partial derivative is called the *delta* of the option, and is easy to view by plotting the Black–Scholes price of an option versus the initial price of the stock and observing its derivative. This is shown in Figure 8.2 for a European call and put option.

Referring to the top plot of Figure 8.2, the delta of a call option starts at 0 and rises eventually to a value of 1, while in the bottom plot, a put option delta starts at -1 and rises to a value of 0. This should correspond to your intuition regarding how many shares of the stock you would hold to hedge the option. For example, if a call option is out-of-the-money and $S << K$ then you don't need to hold the stock to hedge. That is, for a far out-of-the-money call option, the delta of the option should be approximately 0. On the other hand, if the call option is deep in-the-money then it will likely finish in-the-money meaning that the writer of the option will need to deliver the shares. In this case, you will need to have purchased a share of the stock, so delta should be close to 1. Similar intuition holds for the put option.

8.1.3.3 Hedging in Bonds

The hedging concepts can just as easily be applied in the context of interest rates and bonds. Consider the single factor short rate models of Chapter 7, Section 7.3.2 where the short rate follows the process

$$dr_0 = a\,dt + b\,dz$$

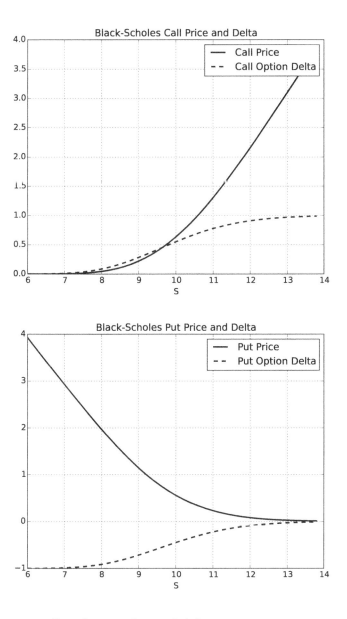

Figure 8.2 European call and put price and delta.

and a and b can be functions of r_0 and t. Following this model, let's create a tradable table that includes two zero-coupon bonds, $B^1(T_1)$ and $B^2(T_2)$, with maturities, T_1 and T_2, and the money market account B_0. We will place bond $B^2(T_2)$ below the horizontal dashes in the tradable table to indicate that it is what we would like to hedge, whereas B_0 and $B^1(T_1)$ are above the dashes since they are what we can hedge with. The augmented tradable table is

Holdings   Prices   Value Change Factor Models

$$- - - \mid - - - - - \mid - - - - - - - - - - - - - -$$

$$\begin{bmatrix} y_0 \\ \hat{y} \\ -- \\ 1 \end{bmatrix} \quad \begin{bmatrix} B_0 \\ B^1(T_1) \\ -- \\ B^2(T_2) \end{bmatrix} \quad d \begin{bmatrix} B_0 \\ B^1(T_1) \\ -- \\ B^2(T_2) \end{bmatrix} = \begin{bmatrix} r_0 B_0 \\ \mathcal{L}B^1(T_1) \\ -- \\ \mathcal{L}B^2(T_2) \end{bmatrix} dt + \begin{bmatrix} 0 \\ bB_r^1(T_1) \\ -- \\ bB_r^2(T_2) \end{bmatrix} dz$$

where

$$\mathcal{L}B^i = (B_t^i + aB_r^i + \frac{1}{2}b^2 B_{rr}^i), \quad i = 1, 2.$$

Now, this portfolio is hedged if $y^T \mathcal{B} = 0$, which leads to the condition

$$\hat{y}bB_r^1(T_1) + bB_r^2(T_2) = 0.$$

Solving for $\hat{y}$ gives

$$\hat{y} = -\frac{B_r^2(T_2)}{B_r^1(T_1)}. \tag{8.18}$$

Note that hedged portfolios of bonds are sometimes called "immunized." (See, for example, [36] or [18].)

Example 8.2 (Computing a Hedge under Vasicek)
Consider hedging a 5-year zero-coupon bond with a 1-year zero-coupon bond when the term structure is driven by a single factor short rate model that follows Vasicek dynamics

$$dr_0 = \kappa(\theta - r_0)dt + bdz$$

with parameters $\kappa = 0.25$, $\theta = 0.07$, and $b = 0.02$ with market price of risk $\lambda = -0.2$. (See Chapter 7, Section 7.3.2.1.) Assume the current value of the short rate is $r_0(0) = 0.05$.

    Let $B^2(5)$ denote the 5-year bond and $B^1(1)$ be the 1-year bond. According to our result in equation (8.18) above, if we hold $B^2(5)$ and would like to hedge it with $B^1(1)$, then we need to calculate the hedging amount $\hat{y}$ as

$$\hat{y} = -\frac{B_r^2(5)}{B_r^1(1)}.$$

To calculate this quantity, let's first recall that the price of a bond under the Vasicek model is given by

$$B(r_0, t|T) = e^{x(t)r_0 + y(t)},$$

where $x(t)$ and $y(t)$ are supplied in equations (7.27) and (7.28) of Chapter 7, Section 7.3.2.1. Thus, the partial derivative with respect to r_0 is given by

$$B_r(r_0, t|T) = x(t)e^{x(t)r_0 + y(t)}.$$

Therefore, for the 1-year and 5-year bonds, using the current value of the short rate as $r_0(0) = 0.05$ and current time $t = 0$, we can calculate the required partial derivatives with respect to r_0 as

$$B_r^1(0.05, 0|1) = -0.83821, \quad B_r^2(0.05, 0|5) = -2.06487.$$

Substituting these values into the formula for the hedging position gives

$$\hat{y} = -\frac{B_r^2(0.05, 0|5)}{B_r^1(0.05, 0|1)} = -\frac{-2.06487}{-0.83821} = -2.4634.$$

Thus, at the current short rate and time, we need to be holding $\hat{y} = 2.4634$ units of the 1-year bond in order to hedge the risk in the 5-year bond.

8.1.3.4 Pricing from the Bond Hedge

Using the same procedure as in the Black–Scholes case, we can convert this hedged portfolio into a potential arbitrage portfolio and derive a pricing equation. Again, we choose y_0 corresponding to the money market account B_0 so that the total cost is 0:

$$y_0 B_0 - \frac{B_r^2(T_2)}{B_r^1(T_1)} B^1(T_1) + B^2(T_2) = 0. \tag{8.19}$$

Note that for no arbitrage to exist, we must have no profit/loss:

$$y_0 r_0 B_0 - \frac{B_r^2(T_2)}{B_r^1(T_1)} \mathcal{L} B^1(T_1) + \mathcal{L} B^2(T_2) = 0. \tag{8.20}$$

Using $y_0 B_0 = \frac{B_r^2(T_2)}{B_r^1(T_1)} B^1(T_1) - B^2(T_2)$ from (8.19) and substituting this into the first term in (8.20) gives

$$r_0 \left(\frac{B_r^2(T_2)}{B_r^1(T_1)} B^1(T_1) - B^2(T_2) \right) - \frac{B_r^2(T_2)}{B_r^1(T_1)} \mathcal{L} B^1(T_1) + \mathcal{L} B^2(T_2) = 0 \tag{8.21}$$

which can be rearranged as

$$\frac{\mathcal{L} B^2(T_2) - r_0 B^2(T_2)}{B_r^2(T_2)} = \frac{\mathcal{L} B^1(T_1) - r_0 B^1(T_1)}{B_r^1(T_1)}. \tag{8.22}$$

Now, since the left side depends only on B^2 and the right side only on B^1, the ratios in equation (8.22) must equal a constant. Let's call that constant λ'. Then by substituting for the definition of $\mathcal{L} B$, the following equation must hold,

$$\frac{B_t^1(T_1) + a B_r^1(T_1) + \frac{1}{2} b^2 B_{rr}^1(T_1) - r_0 B^1(T_1)}{B_r^1(T_1)} = \lambda'. \tag{8.23}$$

To make this look like our previously derived bond pricing equation in Chapter 7, Section 7.3.2, define

$$\lambda' = \lambda b, \tag{8.24}$$

where b is the standard deviation parameter of the short rate process and thus does not change depending on what bond is being considered. This allows (8.23) to be written as

$$\frac{B_t^1(T_1) + aB_r^1(T_1) + \frac{1}{2}b^2 B_{rr}^1(T_1) - r_0 B^1(T_1)}{B_r^1(T_1)} = \lambda b. \tag{8.25}$$

Upon rearranging we have

$$B_t^1(T_1) + (a - \lambda b)B_r^1(T_1) + \frac{1}{2}b^2 B_{rr}^1(T_1) = r_0 B^1(T_1), \tag{8.26}$$

which is the absence of arbitrage bond pricing equation where λ is the market price of risk for dz.

The intuitive explanation of this derivation is that first we hedge out the factor risk by the choice of $\hat{y}$. Since this hedged portfolio has no factor risk, it must earn the same return as the tradable with no factor risk, which is the money market account. This is where the absence of arbitrage pricing equation comes from!

8.1.4 Hedging under Incompleteness

In some cases it is impossible to eliminate all of the factor risk (this is true in *incomplete markets*, see Chapter 5, Section 5.6.1). However, we may still attempt to reduce the effect of the factors on our portfolio, but this will lead to choices in terms of how best to do so.

Here, I present a possible hedge under the jump diffusion model that appears in [53]. Recall the jump diffusion model of Chapter 6, Section 6.6, where we had the following tradable table which is now augmented with the holdings,

Holdings   Prices   Value Change Factor Models

$$\begin{bmatrix} y_0 \\ \hat{y} \\ -- \\ 1 \end{bmatrix} \quad \begin{bmatrix} B \\ S \\ -- \\ c \end{bmatrix} \quad d\begin{bmatrix} B \\ S \\ -- \\ c \end{bmatrix} = \begin{bmatrix} r_0 B \\ (\mu + \alpha\mathbb{E}[Y-1])S \\ --- \\ \mathcal{L}c \end{bmatrix} dt + \begin{bmatrix} 0 & 0 & 0 \\ \sigma S & S & 0 \\ -- & -- & -- \\ \sigma S c_S & 0 & 1 \end{bmatrix} \begin{bmatrix} dz \\ d\psi_1 \\ d\psi_2 \end{bmatrix}$$

and we have suppressed all the left limit notation (such as S^-) for convenience. As a reminder, α is the intensity of the Poisson factor and $\mathcal{L}$ is the operator corresponding to the drift term in Ito's lemma. Moreover, the factors $d\psi_1$ and $d\psi_2$ were defined as

$$\begin{aligned} d\psi_1 &= (Y-1)d\pi(\alpha) - \alpha\mathbb{E}[Y-1]dt, \\ d\psi_2 &= ((c(YS) - c(S))d\pi(\alpha) - \alpha\mathbb{E}[(c(YS) - c(S))]dt). \end{aligned}$$

Note that I have assumed that we hold one unit of the derivative c and will hedge with the stock S. Since the bond B is not driven by any of the risky factors, I do not need to consider it in the hedge. So, the hedged portfolio takes the form

$$
\begin{aligned}
d\mathcal{V}_h &= \hat{y}dS + dc \\
&= \hat{y}((\mu + \alpha\mathbb{E}[Y-1])Sdt + \sigma Sdz + Sd\psi_1) + (\mathcal{L}cdt + \sigma Sc_S dz + d\psi_2) \\
&= [\hat{y}(\mu + \alpha\mathbb{E}[Y-1])S + \mathcal{L}c]\,dt + [\hat{y} + c_S]\sigma Sdz + \hat{y}Sd\psi_1 + d\psi_2. \quad (8.27)
\end{aligned}
$$

Clearly, it is not possible to eliminate all the factor risk by choosing $\hat{y}$. Thus, we are in an incomplete market.

However, we can select $\hat{y}$ to eliminate *some* of the risk over the next dt. In fact, what Merton [40] did in his jump diffusion pricing formula is equivalent to eliminating just the Brownian risk dz. In that case, we would choose

$$
\hat{y} = -c_S \quad (8.28)
$$

where c_S is the partial derivative of Merton's pricing formula given in Section 6.6.2 of Chapter 6.

However, other alternatives are possible. For example, let's choose $\hat{y}$ to minimize the variance of the portfolio over the next dt. Using equation (8.27), one can compute the variance of the change in the hedged portfolio over the next dt as

$$
Var(d\mathcal{V}_h) = [\hat{y} + c_S]^2\, \sigma^2 S^2 dt + \hat{y}^2 S^2 Var(d\psi_1) + Var(d\psi_2) + 2\hat{y}SCov(d\psi_1, d\psi_2).
$$

Minimizing this by setting the derivative with respect to $\hat{y}$ equal to zero and solving for $\hat{y}$ leads to

$$
\hat{y} = -\frac{SCov(d\psi_1, d\psi_2) + c_S\sigma^2 S^2 dt}{S^2 Var(d\psi_1) + \sigma^2 S^2 dt}. \quad (8.29)
$$

To express $\hat{y}$ in a more intuitive form, let's define S^J and c^J to be the changes in S and c, respectively, caused by the jump Y:

$$
S^J = (Y-1)S, \quad c^J = c(YS) - c(S).
$$

In terms of these quantities, one can then calculate that

$$
\begin{aligned}
SCov(d\psi_1, d\psi_2) &= \mathbb{E}[S^J c^J]\alpha dt + Cov(S^J, c^J)\alpha^2 dt^2, \\
S^2 Var(d\psi_1) &= \mathbb{E}[(S^J)^2]\alpha dt + Var(S^J)\alpha^2 dt^2.
\end{aligned}
$$

Finally, substituting these in equation (8.29) and and sending $dt \to 0$ results in

$$
\hat{y} = -\frac{\mathbb{E}[S^J c^J]\alpha + c_S\sigma^2 S^2}{\mathbb{E}[(S^J)^2]\alpha + \sigma^2 S^2}. \quad (8.30)
$$

By writing the hedging amount $\hat{y}$ in this form, we can see that if there are no jumps (i.e., S^J and c^J are zero), it reduces to the standard delta hedge $\hat{y} = -c_S$. Thus, the terms involving S^J and c^J serve the purpose of incorporating the jump risk into the hedge. Overall, this hedge attempts to minimize the combined effects of both the jump and Brownian factors, but doesn't completely eliminate either one.

As you can imagine, there are many other possible hedges for the jump diffusion model. The interested reader is referred to [53]. Thus, in an incomplete market, there is no unique hedge that will reduce all the factor risk.

8.1.5 A Question of Consistency

Note that the hedges derived above often involve knowledge of a pricing formula for the option c. This can be seen by noting that y is often a function of c_S. Thus, our hedging analysis has *presupposed* that the derivative c follows a specified formula that we know.

This *presupposition* of knowledge can bring up a question of consistency between the hedge and the assumption of the formula for c. To illustrate this point, consider the stochastic volatility model from Chapter 6, Section 6.8,

$$
\begin{aligned}
dB &= r_0 B dt & (8.31) \\
dS &= \mu S dt + \sqrt{v} S dz_1 & (8.32) \\
dv &= a dt + b dz_2 & (8.33)
\end{aligned}
$$

with the augmented tradable table

Holdings   Prices   Value Change Factor Models

$$
\begin{bmatrix} y_0 \\ \hat{y} \\ - \\ 1 \end{bmatrix}
\begin{bmatrix} B \\ S \\ - \\ c \end{bmatrix}
\quad d
\begin{bmatrix} B \\ S \\ - \\ c \end{bmatrix} =
\begin{bmatrix} r_0 B \\ \mu S \\ -- \\ \mathcal{L}c \end{bmatrix} dt +
\begin{bmatrix} 0 & 0 \\ \sqrt{v}S & 0 \\ -- & -- \\ \sqrt{v}Sc_S & bc_v \end{bmatrix}
\begin{bmatrix} dz_1 \\ dz_2 \end{bmatrix}
$$

with $\mathcal{L}c = c_t + \mu S c_S + a c_v + \frac{1}{2} v S^2 c_{SS} + \frac{1}{2} b^2 c_{SS} + \rho b \sqrt{v} S c_{Sv}$. Recall that v is the instantaneous variance, which follows its own stochastic differential equation (8.33) driven by dz_2 with correlation coefficient ρ with dz_1.

This is an incomplete market, so we cannot perfectly hedge away the risk. Nevertheless, let's consider a hedged portfolio (with $y_0 = 0$),

$$
\begin{aligned}
d\mathcal{V}_h &= \hat{y}(\mu S dt + \sqrt{v} S dz_1) \\
&\quad + (c_t + \mu S c_S + a c_v + \frac{1}{2} v S^2 c_{SS} + \frac{1}{2} b^2 c_{vv} + \rho b \sqrt{v} S c_{Sv}) dt \\
&\qquad\qquad + \sqrt{v} S c_S dz_1 + b c_v dz_2.
\end{aligned}
$$

Clearly, with $\hat{y}$, we cannot eliminate all the risk. Instead, let's consider a simple hedge where we only eliminate the dz_1 risk. Thus, we choose $\hat{y}\sqrt{v}S + \sqrt{v}Sc_S = 0$ or $\hat{y} = -c_S$. That is, our hedge is to hold $-c_S$ shares of the stock. But what is c_S?

How about using the Black–Scholes formula for $c(S, t)$ and computing c_S from that? But, that is not consistent because to derive the Black–Scholes formula we assume that volatility is *not* stochastic. In our setting above, we are assuming that volatility is stochastic and thus there is no reason to believe that c would follow the Black–Scholes equation. This is where the inconsistency in hedging often arises.

So, you ask, what formula should I use for c? The answer is that you should use the formula that best corresponds to the actual price and movements of c in the market. Thus, if there is stochastic volatility, then you should use a model that you believe most accurately captures that stochastic volatility and how it is reflected in the movement of the derivative c. For incomplete markets, this means that you will likely have to make decisions as to how to value the incompleteness via choices of market prices of risk and the like. Unfortunately, this answer is not completely satisfying, but reflects the basic truth that absence of arbitrage considerations alone don't provide enough information to price (and hedge) uniquely in incomplete markets.

8.2 HEDGING FROM AN UNDERLYING VARIABLE PERSPECTIVE

In some cases, there is a simpler and faster route to derive hedges that doesn't involve an explicit use of the factors. In this approach, we consider the asset that we are trying to hedge to be a function of underlying variables, and hedge by eliminating the sensitivity to those underlying variables. In the simplest terms, sensitivity is measured by the (calculus-based) derivative of the hedged portfolio with respect to the underlying variables of concern. Figure 8.3 shows how this perspective fits into our modeling paradigm.

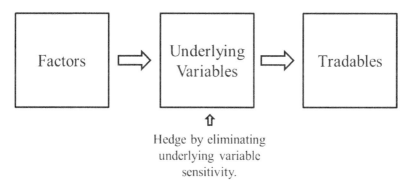

Figure 8.3 Hedging from an underlying variable sensitivity perspective.

8.2.1 Black–Scholes Hedging

For example, in the Black–Scholes setup, we assume that an option is a function of the underlying stock S and time t, which we write as $c(S,t)$. Now, if we would like to hedge out the risk in our option, we note that all the risk in the price of an option comes from its dependence on the stock variable. Hence, we form a hedged portfolio

$$V_h = c(S,t) + yS \tag{8.34}$$

and get to choose $\hat{y}$. However, since all the risk comes from the stock variable, we will be hedged if our portfolio has no sensitivity to changes in the stock. Another way of saying this is that we want the (calculus-based) *derivative* of V_h with respect to S to be 0. That means that to first order, small changes in S will cause no change in the value of the hedged portfolio V_h. This is what it means to be hedged!

This condition can be written as

$$\frac{\partial \mathcal{V}_h}{\partial S} = c_S + \hat{y} = 0 \tag{8.35}$$

Solving for $\hat{y}$ yields $\hat{y} = -c_S$ which is the same answer that we arrived at using factors. As a reminder, this is known as a *delta* hedge and the quantity c_S is commonly referred to as the delta of the option.

Example 8.3 (Plotting the Hedged Portfolio)
There is a simple graphical interpretation of a delta hedge for an option. Consider a plot with the underlying variable on the x-axis, and the value of the tradables as a function of the underlying variable on the y-axis. In this example, the underlying variable is the stock price S and the tradables are the stock S and a European call option $c(S, t)$ following the Black-Scholes formula with current time $t = 0$, strike $K = 10$, and expiration $T = 0.25$. The volatility of the stock is assumed to be $\sigma = 0.3$, while the risk-free rate is taken as $r_0 = 0.03$. The top plot of Figure 8.4 shows the stock and call values as a function of the price of the stock S.

Now, assume that the current price of the stock is $S = 10$ and a delta hedged portfolio is formed containing one option c and $\hat{y} = -c_S(S, t)$ shares of the stock where $\hat{y}$ is evaluated at $t = 0$ and $S = 10$. The bottom plot of Figure 8.4 shows the overall value of this delta hedged portfolio as a function of S. As can be seen in the plot, the delta hedged portfolio value $\mathcal{V}_h = \hat{y}S + c$ has derivative equal to zero at $S = 10$. That is, the delta hedge combines the stock and option values given in the top plot to form an overall portfolio whose derivative is zero at the current price of the stock. This is the essence of a delta hedge.

8.2.2 Hedging Bonds

We can apply this same underlying variables approach to our hedging of bonds. In that case, we hold a bond $B^2(r_0, t|T_2)$ that is a function of the short rate $r_0(t)$ and time t. We wish to hedge this bond with another bond $B^1(r_0, t|T_1)$ that is also a function of the short rate $r_0(t)$ and time t. In this setup, the risk in the price of a bond comes from the short rate $r_0(t)$ which is our underlying variable. Thus, a portfolio $\mathcal{V}_h = B^2(r_0, t|T_2) + \hat{y}B^1(r_0, t|T_1)$ is hedged if its derivative with respect to r_0 is zero:

$$\frac{\partial \mathcal{V}_h}{\partial r} = B_r^2(T_2) + \hat{y}B_r^1(T_1) = 0. \tag{8.36}$$

Solving for $\hat{y}$ yields $\hat{y} = -\frac{B_r^2(T_2)}{B_r^1(T_1)}$ which is the same answer that we found using factors.

8.2.3 Derivatives Imply Small Changes

Note that this underlying variable approach uses the calculus-based derivative as a measure of sensitivity. Recall that the derivative is the change in a function for a

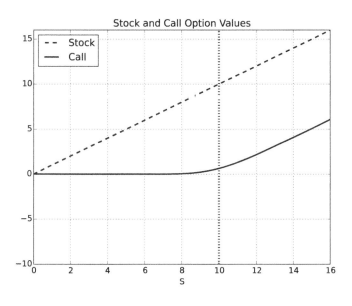

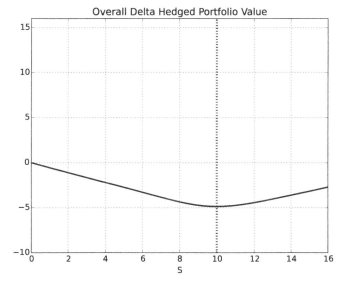

Figure 8.4 Delta hedge: Top: Plot of the stock and call option, Bottom: Plot of the hedged portfolio assuming that the current price of the stock is $10.

small change in the variable. Thus, when we say that a portfolio is hedged against stock price movements because its derivative with respect to the stock is zero, this means the value of the portfolio change is approximately 0 for *small* changes in the price of the stock. However, the value of the portfolio change could be quite large if the stock change is large. Hence, this approach works only if we know that the stock price changes (or equivalently, the underlying variable changes) will be small over the next dt. This is the case for stock price movements driven by Brownian motion since Brownian motion paths are continuous. However, if the stock price is driven by a Poisson process, then we would expect large jumps at times. Thus, in this case using the derivative is not a good approach since the stock price change would be large and the derivative would likely not be a good approximation to the change in the portfolio value. This was the situation in the jump diffusion model in Section 8.1.4, and note that the hedging strategy derived in (8.30) was not the same one that would result from this underlying variable derivative-based approach.

8.3 HIGHER ORDER APPROXIMATIONS

Using the derivative to model the change in a portfolio due to a change in a variable is a linear approximation of the portfolio value as a function of the underlying variable. Of course, higher order approximations to the portfolio value can also be used. An easy way to do this is to use the terms of a Taylor expansion. This leads to the so-called Greeks.

8.3.1 The Greeks

In general, a call option is not just a function of the price of the stock and time. From the Black–Scholes formula, we see that it depends also on time t, the risk-free rate r_0, and the volatility σ. In our derivation of the Black–Scholes formula, we assumed that r_0 and σ were constant (not driven by random factors), and thus we did not need to explicitly consider the dependence on these variables in our calculations. However, recognizing that this is only an approximation, we can allow them to change and then ask how their changes might cause the price of an option to change assuming that the price follows the Black–Scholes formula (which is not completely consistent!). To do this, we could just construct a multivariable Taylor series expansion of, for example, the Black–Scholes formula for a European call option from equation (6.7) as

$$\Delta c = c_t \Delta t + c_S \Delta S + c_r \Delta r_0 + c_\sigma \Delta \sigma + \frac{1}{2} c_{SS} (\Delta S)^2 + \dots \qquad (8.37)$$

I have included only a single second order term (there are many second order terms), but you are welcome to extend the expansion as far as you like. The terms covered here are the ones that are "traditionally" analyzed.

By looking at this Taylor expansion, we see that the various derivatives tell us the sensitivity of the price of an option to changes in those variables. Each of these derivatives is given a Greek letter name (or Greek "sounding" letter such as "vega") as shown below.

Greek	Definition
delta (Δ)	$c_S = \frac{\partial c}{\partial S}$
gamma (Γ)	$c_{SS} = \frac{\partial^2 c}{\partial S^2}$
theta (Θ)	$c_t = \frac{\partial c}{\partial t}$
rho (ρ)	$c_{r_0} = \frac{\partial c}{\partial r_0}$
vega	$c_\sigma = \frac{\partial c}{\partial \sigma}$

If you have a poor memory, like I do, then you can remember some of the Greeks by noting that the first letter of the Greek is the same as the partial derivative. That is, "theta" starts with "t" and is the partial derivative with respect to t. Similarly, "rho" starts with "r" and is with respect to r_0, and "vega" (which is not a real Greek letter, but perhaps sounds like one) starts with "v" and is the partial with respect to "volatility". Unfortunately, with "delta" and "gamma" you are on your own.

Note that in the typical dynamic hedging assumption, we are able to trade continuously. Thus, only terms of order dt and lower matter. However, this Taylor expansion approach assumes that we are not trading continuously. Thus, we use a Δt instead of a dt and furthermore, higher order terms will enter. In many ways, this is actually more practical than the typical continuous time trading assumptions.

The following example provides formulas for the Greeks associated with the Black–Scholes formula.

Example 8.4 (Greeks for Black–Scholes Calls and Puts)
Consider the Black–Scholes(–Merton) model from Chapter 6, Section 6.2 for a stock paying a continuous dividend at a rate of q. Recall that the prices of European call and put options are

$$
\begin{aligned}
c &= Se^{-q(T-t)}\Phi(d_1) - Ke^{-r_0(T-t)}\Phi(d_2), \\
p &= Ke^{-r_0(T-t)}\Phi(-d_2) - Se^{-q(T-t)}\Phi(-d_1)
\end{aligned}
$$

where

$$
d_1 = \frac{\ln(S/K) + (r_0 - q + \frac{1}{2}\sigma^2)(T-t)}{\sigma\sqrt{T-t}}, \tag{8.38}
$$

$$
d_2 = d_1 - \sigma\sqrt{T-t}. \tag{8.39}
$$

and $\Phi(x)$ is the cumulative distribution function of the standard normal random variable with mean 0 and standard deviation 1. Its derivative is $\Phi'(x) = \frac{1}{\sqrt{2\pi}}e^{-\frac{x^2}{2}}$.

Under this model, the formulas for the Greeks for European calls and puts are

Quantity	Call	Put
Black–Scholes price	c	p
delta ($\Delta = \frac{\partial c}{\partial S}$)	$e^{-q(T-t)}\Phi(d_1)$	$-e^{-q(T-t)}\Phi(-d_1)$
gamma ($\Gamma = \frac{\partial^2 c}{\partial S^2}$)	$\frac{\Phi'(d_1)e^{-q(T-t)}}{S\sigma\sqrt{T-t}}$	$\frac{\Phi'(d_1)e^{-q(T-t)}}{S\sigma\sqrt{T-t}}$
theta ($\Theta = \frac{\partial c}{\partial t}$)	$r_0 c - (r_0 - q)S\Delta - \frac{1}{2}\sigma^2 S^2\Gamma$	$r_0 p - (r_0 - q)S\Delta - \frac{1}{2}\sigma^2 S^2\Gamma$
rho ($\rho = \frac{\partial c}{\partial r_0}$)	$K(T-t)e^{-r_0(T-t)}\Phi(d_2)$	$-K(T-t)e^{-r_0(T-t)}\Phi(-d_2)$
vega ($\frac{\partial c}{\partial \sigma}$)	$S(\sqrt{T-t})\Phi'(d_1)e^{-q(T-t)}$	$S(\sqrt{T-t})\Phi'(d_1)e^{-q(T-t)}$

Note that I have written the formula for theta in terms of the price, delta, and gamma of the corresponding call or put option. One can write an explicit formula for theta by substituting for the price, delta, and gamma, but it looks much messier.

As an example of these calculations, consider the following parameter values:

$$S = 10, \ K = 9, \ \sigma = 0.3, \ r_0 = 5\%, \ T = 0.25, \ q = 1\%,$$

then the values of the Greeks for European call and put options are given by

Quantity	Call	Put
Black–Scholes price	1.265769691	0.178938671
delta ($\Delta = \frac{\partial c}{\partial S}$)	0.798685672	-0.198817451
gamma ($\Gamma = \frac{\partial^2 c}{\partial S^2}$)	0.185791387	0.185791387
theta ($\Theta = \frac{\partial c}{\partial t}$)	-1.092247024	-0.747587326
rho ($\rho = \frac{\partial c}{\partial r_0}$)	1.680271757	-0.541778294
vega ($\frac{\partial c}{\partial \sigma}$)	1.3934354	1.3934354.

8.3.2 Delta–Gamma Hedge

We can think of a delta hedge where $\hat{y} = -c_S$, as in Example 8.2.1, as eliminating the first order term for ΔS in the Taylor expansion of (8.37). Of course, we can go further and ask whether higher order terms can be eliminated as well.

If you eliminate both the first order term in ΔS and the second order term $(\Delta S)^2$, this is called a *delta–gamma* hedge. For such a delta–gamma hedge you need more than just the underlying stock, and furthermore, you need a tradable whose Taylor expansion depends on the second order term $(\Delta S)^2$. Let's demonstrate how this works by using a second call option $c^{(2)}$ and the underlying stock to delta–gamma hedge an option c. In this case, the hedged portfolio is

$$\mathcal{V}_h = c + \hat{y}_1 S + \hat{y}_2 c^{(2)},$$

which can be Taylor expanded to obtain

$$
\begin{aligned}
\Delta \mathcal{V}_h(S,t) &= \Delta c + \hat{y}_1 \Delta S + \hat{y}_2 \Delta c^{(2)} \\
&= \left[c_t \Delta t + c_S \Delta S + \frac{1}{2} c_{SS} (\Delta S)^2 \right] + \hat{y}_1 \Delta S \\
&\qquad\qquad + \hat{y}_2 \left[c_t^{(2)} \Delta t + c_S^{(2)} \Delta S + \frac{1}{2} c_{SS}^{(2)} (\Delta S)^2 \right] \\
&= \left[c_t + \hat{y}_2 c_t^{(2)} \right] \Delta t + \left[c_S + \hat{y}_1 + \hat{y}_2 c_S^{(2)} \right] \Delta S + \frac{1}{2} \left[c_{SS} + \hat{y}_2 c_{SS}^{(2)} \right] (\Delta S)^2.
\end{aligned}
$$

To delta–gamma hedge, we have to eliminate the coefficients of ΔS and $(\Delta S)^2$ by choosing $\hat{y}_1$ and $\hat{y}_2$. That is, we must solve the two equations:

$$
\begin{aligned}
c_S + \hat{y}_1 + \hat{y}_2 c_S^{(2)} &= 0, \\
c_{SS} + \hat{y}_2 c_{SS}^{(2)} &= 0.
\end{aligned}
$$

The solution is

$$
\hat{y}_1 = \frac{c_{SS}}{c_{SS}^{(2)}} c_S^{(2)} - c_S, \qquad \hat{y}_2 = -\frac{c_{SS}}{c_{SS}^{(2)}}, \tag{8.40}
$$

which provides us with the formulas for creating a delta–gamma hedge.

Example 8.5 (Delta–Gamma Hedge)
Consider a stock with current price $S = \$10$ and volatility $\sigma = 0.3$. Assume that there are two different European call options written on the stock. The first option, denoted c, has strike price $K = \$9.5$ with expiration $T = 0.25$, while the second option, denoted $c^{(2)}$, has strike $K_2 = \$8.5$ and expiration $T_2 = 0.4$. The risk-free rate is $r_0 = 0.03$.

Assume that we hold the first option c and would like to delta–gamma hedge our position using the stock S and the second option $c^{(2)}$. According to (8.40), we should hold $\hat{y}_1$ shares of the stock and $\hat{y}_2$ of the second option, where $\hat{y}_1$ and $\hat{y}_2$ are calculated as

$$
\hat{y}_1 = \frac{c_{SS}}{c_{SS}^{(2)}} c_S^{(2)} - c_S = 0.9237414, \qquad \hat{y}_2 = -\frac{c_{SS}}{c_{SS}^{(2)}} = -1.897903.
$$

Note that the calculation of $\hat{y}_1$ and $\hat{y}_2$ is done at the current value of the stock, $S = \$10$.

The overall hedged portfolio is then given by $c + \hat{y}_1 S + \hat{y}_2 c^{(2)}$, and this hedged position should have both its first and second partial derivatives with respect to S equal to 0 at the current value of the stock, which is $S = \$10$. This is shown in Figure 8.5. The top plot shows the values of the three securities as a function of S, while the bottom plot shows the value of the overall hedged portfolio which has first and second derivative equal to 0 at $S = \$10$.

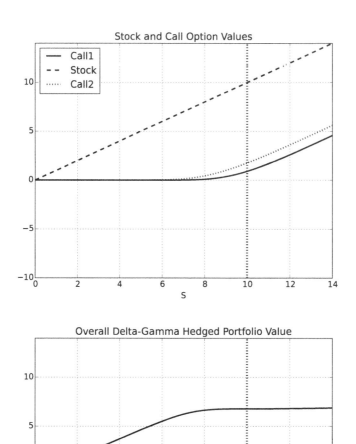

Figure 8.5 Delta–gamma hedge: Top: Plot of the stock and call options, Bottom: Plot of the hedged portfolio assuming that the current price of the stock is $10.

8.3.3 Determining What the Error Looks Like

The underlying variable approach using Taylor series expansions also allows us to look at the terms that are not hedged. In this sense, we can see what the error looks like in various hedges. As an example, let's take a peek at the error in a delta hedge under Black–Scholes assumptions. Consider the hedged portfolio,

$$\mathcal{V}_h(S,t) = c(S,t) - c_S S, \tag{8.41}$$

where I have already plugged in $\hat{y} = -c_S$, and let's perform a Taylor expansion of this in S and t. Note that I could also look at the terms involving r_0, σ, etc., but for now I will just focus on S and t. The Taylor expansion is

$$\Delta \mathcal{V}_h(S,t) = \Delta c(S,t) - c_S \Delta S = c_t \Delta t + \frac{1}{2} c_{SS} (\Delta S)^2 + \ldots \qquad (8.42)$$

where I have kept only terms up to order Δt. Next, we could substitute in the formulas for c_t and c_{SS} using the Black–Scholes formula, but another approach will actually be more revealing.

Assume that the price is following the Black Scholes equation, but under the *implied volatility* σ_i. As a reminder, the implied volatility is the value of σ that makes the Black–Scholes formula equal the market price of the option. (See Chapter 6, Section 6.1.2.) Under this assumption, c satisfies

$$c_t + r_0 S c_S + \frac{1}{2} \sigma_i^2 S^2 c_{SS} = r_0 c. \qquad (8.43)$$

Said another way, this equation assumes that the market is pricing the option using the Black–Scholes formula with a volatility value of σ_i. Thus, rather than using the explicit Black–Scholes formula to calculate c_t, we can instead use (8.43) to substitute for c_t in (8.42), which leads to

$$
\begin{aligned}
\Delta \mathcal{V}_h(S,t) &= \left(r_0 c - r_0 S c_S - \frac{1}{2} \sigma_i^2 S^2 c_{SS} \right) \Delta t + \frac{1}{2} c_{SS} (\Delta S)^2 \\
&= r_0 (c - c_S S) \Delta t + \frac{1}{2} c_{SS} \left((\Delta S)^2 - \sigma_i^2 S^2 \Delta t \right). \qquad (8.44)
\end{aligned}
$$

Finally, noting that $\mathcal{V}_h = c - c_S S$ gives

$$\Delta \mathcal{V}_h(S,t) = r_0 \mathcal{V}_h \Delta t + \frac{1}{2} c_{SS} \left((\Delta S)^2 - \sigma_i^2 S^2 \Delta t \right). \qquad (8.45)$$

We can interpret this equation as follows. The first term $r_0 \mathcal{V}_h \Delta t$ indicates that the rate of return on the hedged portfolio $\mathcal{V}_h$ is the risk-free rate r_0. This makes sense because a hedged portfolio should replicate a risk-free security since the purpose of hedging is to remove the risk.

The second term $\frac{1}{2} c_{SS} \left((\Delta S)^2 - \sigma_i^2 S^2 \Delta t \right)$ indicates that the change in the hedged portfolio will also depend upon the difference between the squared movement of the stock $(\Delta S)^2$, and the implied volatility related term $\sigma_i^2 S^2 \Delta t$. If these two are equal then this term will vanish. However, if they are not equal, this will contribute to the change in the value of the hedged portfolio.

To understand this term a little better, let's approximate the stochastic differential equation factor model for S as $\Delta S = \mu S \Delta t + \sigma S \Delta z$, and note that this implies that

$$
\begin{aligned}
(\Delta S)^2 &= (\mu S)(\Delta t)^2 + 2\mu\sigma S^2 \Delta t \Delta z + (\sigma S)^2 (\Delta z)^2 \\
&\approx (\mu S)(\Delta t)^2 + 2\mu\sigma S^2 \Delta t \Delta z + (\sigma S)^2 \Delta t
\end{aligned}
$$

where the approximation of $(\Delta z)^2 \approx \Delta t$ is related to our derivation of Ito's lemma. (See Chapter 2, Section 2.2.1.) The upshot is that to order Δt, we have

$$(\Delta S)^2 \approx (\sigma S)^2 \Delta t.$$

Substituting this into equation (8.44) leads to

$$\Delta \mathcal{V}_h(S,t) = r_0 \mathcal{V}_h \Delta t + \frac{1}{2} c_{SS} \left(\left(\sigma^2 - \sigma_i^2 \right) S^2 \right) \Delta t. \tag{8.46}$$

Equation (8.46) shows clearly that the second term depends on the difference between the actual volatility of the stock σ versus the implied volatility σ_i. For this hedge where we are long the option and short delta of the stock, since $c_{SS} > 0$ we make money (over the risk-free rate) if the actual volatility of the stock is more than the implied volatility, and vice versa if the actual volatility is less than the implied volatility. In particular, this analysis shows that a delta-hedged option can be used to make a bet on realized volatility σ relative to implied volatility σ_i.

8.4 SUMMARY

Hedging can be approached from two different points of view. From the first point of view, we recognize that risk comes from the factors. Thus, in hedging we try to eliminate the factor risk. This idea is appropriate regardless of what the risky factors are.

In the second point of view, tradables are viewed as functions of underlying variables, and we hedge by constructing a portfolio that eliminates the sensitivity to moves in the underlying variables. Usually we do this by setting to zero the (calculus-based) derivative of the hedged portfolio with respect to the underlying variable. This derivative condition is imposed at the current value of the underlying variable. Since we set only the derivative to zero, and the (calculus-based) derivative reflects a local approximation to the portfolio, this works as long as there are small moves in the underlying variable between rehedging opportunities. Furthermore, if we have more tradables to place in our portfolio, we can set higher order derivatives of our portfolio to zero and create better hedges. This approach is nothing more than eliminating terms in the Taylor expansion of the hedged portfolio. In fact, if you understand Taylor expansions well, then you are on your way to mastering the majority of hedging methods.

EXERCISES

8.1 Consider a European call option with expiration $T = 0.25$ and strike price $K = \$10$ on a non-dividend paying stock with current price $S(0) = \$10.50$ and volatility $\sigma = 35\%$. The risk-free rate is 5%. If you own the call option and would like to hedge your position by holding shares of the stock, how many shares should you be holding?

8.2 Consider a European put option with expiration $T = 0.5$ and strike price $K = \$50$ on a non-dividend paying stock with current price $S(0) = \$51$ and volatility $\sigma = 25\%$. The risk-free rate is 3%. If you own the put option and would like to hedge your position by holding shares of the stock, how many shares should you be holding?

8.3 Consider a European call option with expiration $T = 0.15$ and strike price $K = \$95$ on a stock that pays a continuous dividend at a rate of $q = 2\%$. The stock has current price $S(0) = \$100$ and volatility $\sigma = 40\%$. The risk-free rate is 4%. If you own the call option and would like to hedge your position by holding shares of the stock, how many shares should you be holding?

8.4 Consider a European put option with expiration $T = 0.3$ and strike price $K = \$26$ on a stock that pays a continuous dividend at a rate of $q = 4\%$. The stock has current price $S(0) = \$25$ and volatility $\sigma = 20\%$. The risk-free rate is 7%. If you own the put option and would like to hedge your position by holding shares of the stock, how many shares should you be holding?

8.5 Consider a European call option with expiration $T = 0.3$ and strike price $K = \$100$ on a stock S. The stock has current price $S(0) = \$100$ and volatility $\sigma = 30\%$. The risk-free rate is 5%. Compute the delta of the stock for continuous dividend rates of $q = 0$, 0.02, 0.04, 0.06. Is the delta of the option increasing or decreasing with increasing q? All else being equal, if a stock has a higher dividend rate, would you need to be short more or fewer shares of S to be hedged? Explain your answer.

8.6 Repeat Exercise 8.5 for a European put option.

8.7 Assume that the term structure of interest rates follows Vasicek dynamics with short rate parameters $\kappa = 0.25$, $\theta = 0.07$, and $b = 0.02$. The market price of risk is assumed to be $\lambda = -0.2$ and the current short rate is $r_0(0) = 0.06$. Consider a zero-coupon bond with maturity $T = 2$ years. Calculate the number of bonds of maturity $T = 6$ years that would need to be held to hedge the factor risk in the 2-year bond.

8.8 Assume that the term structure of interest rates follows Vasicek dynamics with $\kappa = 0.6$, $\theta = 0.03$, and $b = 0.015$. The market price of risk is assumed to be $\lambda = -0.1$ and the current short rate is $r_0(0) = 0.05$. Consider a zero-coupon bond with maturity $T = 10$ years. Calculate the number of bonds of maturity $T = 3$ years that would need to be held to hedge the factor risk in the 10-year bond.

8.9 Consider a European call option, c, with strike $K = \$11$ and expiration $T = 0.3$ on a non-dividend paying stock S. Consider delta–gamma hedging this call option with the stock S and another call option $c^{(2)}$ on the stock with strike $K = \$10$ and $T = 0.2$. How many shares of the stock S and options $c^{(2)}$

should be held if the current stock price is $S(0) = \$10$, its volatility is $\sigma = 0.3$, and the risk-free rate is $r_0 = 5\%$?

8.10 Consider a European call option, c, with strike $K = \$52$ and expiration $T = 0.2$ on a non-dividend paying stock S. Consider delta–gamma hedging this call option with the stock S and another call option $c^{(2)}$ on the stock with strike $K = \$48$ and $T = 0.5$. How many shares of the stock S and options $c^{(2)}$ should be held if the current stock price is $S(0) = \$50$, its volatility is $\sigma = 0.4$, and the risk-free rate is $r_0 = 4\%$?

8.11 Compute the Greeks indicated in the following table for a European call option with strike $K = \$24$ and expiration $T = 0.3$ on a non-dividend paying stock with current price $S(0) = \$25$ and volatility $\sigma = 0.25$. The risk-free rate is $r_0 = 0.05$.

Quantity	Call
Black–Scholes price	?
delta $(\Delta = \frac{\partial c}{\partial S})$	?
gamma $(\Gamma = \frac{\partial^2 c}{\partial S^2})$	?
theta $(\Theta = \frac{\partial c}{\partial t})$	?
rho $(\rho = \frac{\partial c}{\partial r_0})$	?
vega $(\frac{\partial c}{\partial \sigma})$	?

8.12 Repeat Exercise 8.11 for a European put option instead of a call option.

8.13 Compute the Greeks indicated in the following table for a European call option with strike $K = \$70$ and expiration $T = 0.15$ on a stock with current price $S(0) = \$68$, volatility $\sigma = 0.3$, and continuous dividend rate $q = 0.02$. The risk-free rate is $r_0 = 0.03$.

Quantity	Call
Black–Scholes price	?
delta $(\Delta = \frac{\partial c}{\partial S})$	?
gamma $(\Gamma = \frac{\partial^2 c}{\partial S^2})$	?
theta $(\Theta = \frac{\partial c}{\partial t})$	?
rho $(\rho = \frac{\partial c}{\partial r_0})$	?
vega $(\frac{\partial c}{\partial \sigma})$	?

8.14 Repeat Exercise 8.13 for a European put option instead of a call option.

CHAPTER 9

Computation of Solutions

T HIS chapter provides an introduction to computational methods used in derivative pricing from the factor model perspective. In particular, we explore the popular binomial tree approach to the pricing of derivative securities [14, 43].

This method is explained by showing that it follows from a discretization of the factor model approach and three step procedure that we used to derive absence of arbitrage equations in previous chapters. In particular, we rely upon the simple binary approximation to a Brownian factor given in Chapter 1. That binary approximation allows us to describe our factor dynamics on a tree that is created by binary up and down moves of the factor. Since underlying variables and tradables are ultimately also driven by the factors, they become representable on the same tree structure. In essence, discretization of the factors induces discretization of the underlying variables and tradables as it flows through the modeling paradigm. Finally, calibration and absence of arbitrage pricing can be performed numerically on the resulting trees, resulting in a systematic computational methodology.

We demonstrate this approach using three increasingly extensive examples. The first example demonstrates how to price both European and American call and put options on a stock that follows geometric Brownian motion. The next example considers calibration and pricing of an interest rate derivative using a CIR based single factor short rate model. The final example implements a single factor LIBOR Market Model, along with a calibration algorithm and the pricing of a simple interest rate derivative.

9.1 DISCRETIZING THE MODELING PARADIGM

Recall our basic modeling paradigm as presented in Figure 9.1. Modeling begins at the factors, which drive the underlying variables through factor models, and then tradables are functions of the underlying variables.

To compute the price of derivatives, the approach we take here begins from a discretization of the factors (and time). This discretization of the factors then flows to the underlying variables and the tradables, and eventually allows us to compute the prices of derivatives. By taking this route, we create an approach

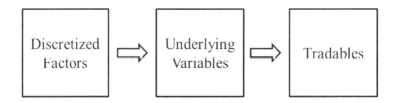

Figure 9.1 Discretized factors drive the modeling paradigm for computation.

that is completely consistent with the paradigm developed in this book. In fact, this chapter basically translates the paradigm of Chapter 5 into a discrete time and factor setting. The advantage of using this discrete setting is that it provides us with a computational methodology to actually solve the resulting absence of arbitrage equations and price derivatives.

To get a feel for this discretized factor model approach, we first show how continuous factor models driven by Brownian factors, dz, can be discretized using a binary approximation. This is the same approximation given in Chapter 1, Section 1.1.5. After seeing this discretization, we then utilize its basic structure to motivate the development of our modeling and computational methods.

9.1.1 Discretizing Factor Models

We begin by discretizing factor models. At the most basic level, this is achieved by discretizing both the random factors and the time factor.

Discretizing the time factor dt is easy. We simply replace dt by Δt which is some "small" but finite amount of time.

On the other hand, there are potentially many options for discretization of the random factors, such as a Brownian motion factor dz. But, one of the simplest is the binary approximation given in Section 1.1.5 of Chapter 1. To review, a simple approximation of a Brownian motion is shown below in Figure 9.2. In this case,

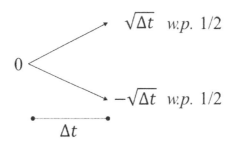

Figure 9.2 Binary model of an increment in Brownian motion.

we replace dz which is Gaussian distributed with mean 0 and variance dt, with a binary random variable approximation $Z\sqrt{\Delta t}$ where Z is the random variable

$$Z = \begin{cases} 1 & w.p. \ \frac{1}{2} \\ -1 & w.p. \ \frac{1}{2} \end{cases}$$

and $w.p.$ stands for "with probability." This binary approximation to the Gaussian, $dz \approx Z\sqrt{\Delta t}$, has mean 0 and variance Δt.

9.1.2 Discretizing Geometric Brownian Motion

Before providing a general discretized modeling paradigm driven by a binary factor, we first demonstrate how the binary approximation can be used to discretize our favorite example of geometric Brownian motion.

9.1.2.1 Direct Discretization of the Factor Model

Perhaps the easiest way to discretize a factor model based on geometric Brownian motion is to directly discretize the factors. That is, if we start with a factor model of the form

$$dS = \mu S dt + \sigma S dz, \tag{9.1}$$

then we simply replace dt by Δt and dz by the binary approximation $Z\sqrt{\Delta t}$ given in Figure 9.2. That is, the discretized version of the factor model is

$$\Delta S = \mu S \Delta t + \sigma S \sqrt{\Delta t} Z, \tag{9.2}$$

with the new factors, Δt and Z. This is still in our standard factor model form, just in discrete time with

$$\Delta S = \mathcal{A}\Delta t + \mathcal{B}Z$$

and

$$\mathcal{A} = \mu S, \quad \mathcal{B} = \sigma S \sqrt{\Delta t}.$$

However, for the continuous time geometric Brownian motion model, we also have a closed form solution that can be used to create an alternative discretized factor model, as shown next.

9.1.2.2 Discretizing the Exact Solution

In the case of geometric Brownian motion as in equation (9.1), we know that the exact closed form solution is

$$S(t) = S(t_0)e^{\left(\left(\mu - \frac{1}{2}\sigma^2\right)(t - t_0) + \sigma(z(t) - z(t_0))\right)}$$

where t_0 is the initial time. If, in this case, we apply this solution over the time increment t_0 to $t_0 + \Delta t$, and then discretize using $z(t_0 + \Delta t) - z(t_0) \approx Z\sqrt{\Delta t}$, we have

$$S(t_0 + \Delta t) = S(t_0)e^{\left(\left(\mu - \frac{1}{2}\sigma^2\right)\Delta t + \sigma\sqrt{\Delta t}Z\right)}$$

which is no longer in the form of a factor model. Instead, it simply provides a model where with equal probability $S(t_0 + \Delta t)$ can take the two values,

$$
S(t_0 + \Delta t) = \begin{cases} S^u(t_0 + \Delta t) = S(t_0)e^{\left((\mu - \frac{1}{2}\sigma^2)\Delta t + \sigma\sqrt{\Delta t}\right)} & w.p. \ \frac{1}{2} \\ S^d(t_0 + \Delta t) = S(t_0)e^{\left((\mu - \frac{1}{2}\sigma^2)\Delta t - \sigma\sqrt{\Delta t}\right)} & w.p. \ \frac{1}{2} \end{cases} \tag{9.3}
$$

where the up value S^u was obtained by replacing Z with its up value of 1, and the down value S^d was obtained by replacing Z with its down value -1. This binary representation for $S(t_0 + \Delta t)$ is given in Figure 9.3.

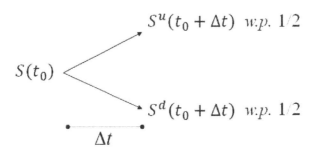

Figure 9.3 Binary approximation for $S(t_0 + \Delta t)$.

9.1.3   More General Modeling Induced by Binary Factor

The form of the model above for geometric Brownian motion in equation (9.3) and Figure 9.3 suggests a more general binary structure induced by using a binary model for the factor. Over each time step Δt, the binary random factor Z can take the value 1 with probability p or -1 with probability $1 - p$. We generically call $Z = 1$ the *up* state of the factor, and $Z = -1$ the *down* state.

For simplicity, let's call the underlying variable S. Then, when the factor moves to the up state, we denote the value of the underlying variable S^u, and when the factor moves to the down state, we denote the value of the underlying variable as S^d.

Moreover, consider a tradable that is a function of the underlying variable S, which we denote $c(S, t)$. Suppressing the time argument t for the moment, we see that the binary model for S also induces a binary model for the tradable $c(S)$ via

$$ c^u = c(S^u), \quad c^d = c(S^d). $$

The flow of the binary structure from the factor to the underlying variable and finally to the tradable is depicted in the context of our modeling paradigm in Figure 9.4.

9.1.4   From the Binary Model to a Factor Model

Before proceeding to the three step procedure to derive the absence of arbitrage equations, we need to be able to represent the underlying variables and tradables

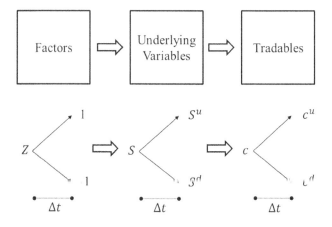

Figure 9.4 Flow of binary modeling from factors to underlying variables to tradables.

as factor models. Using the models induced by the binary random factor, as shown in Figure 9.4, we can convert these to factor models driven by Z by noting that

$$\Delta S = S(t_0 + \Delta t) - S(t_0) = \frac{1}{2} \left(S^u + S^d \right) - S + \frac{1}{2} \left(S^u - S^d \right) Z.$$

where, notationally, we will often drop the time argument of variables (as done on the right side of this equation), with the understanding that S without a superscript represents that value at the *beginning* of the time period of length Δt, and S^u and S^d represent the up and down values at the *end* of the time period.

You should verify that the above equation indeed corresponds to the movement of S in the binary model. Moreover, to expose the explicit factor model structure, we can write this as

$$\Delta S = \mathcal{A}\Delta t + \mathcal{B}Z \tag{9.4}$$

where

$$\mathcal{A} = \frac{1}{2\Delta t} \left(S^u + S^d - 2S \right), \quad \mathcal{B} = \frac{1}{2} \left(S^u - S^d \right). \tag{9.5}$$

The exact same approach and equations work for creating a factor model for the derivative c, or for any variable defined on the binary structure. That is, these equations are the generic formula for turning any binary induced process into a corresponding discrete factor model. We will return to them later.

9.1.4.1 Note on Time Scaling of the Factor

A direct discretizing of the Brownian factor dz with time dt would define $\Delta z \approx Z\sqrt{\Delta t}$ as the discretized factor. We have chosen to use Z as the factor, without the $\sqrt{\Delta t}$ scaling.

The advantage of choosing $\Delta z = Z\sqrt{\Delta t}$ as the factor is that it scales properly with time. Furthermore, the market price of risk corresponding to Δz will be an annualized market price of risk and comparable regardless of the size of the time step Δt. On the other hand, using the factor Z which is always 1, -1 regardless of the time step leads to a market price of risk that does not scale with time. However, it is easy enough to annualize the market price of risk for Z to correspond to a market price of risk for $\Delta z = Z\sqrt{\Delta t}$ by simply multiplying by $\sqrt{\Delta t}$. (You are asked to verify this in Exercise 9.6.) We will use Z as the discrete binary factor throughout this chapter.

9.2   THREE STEP PROCEDURE UNDER BINARY MODELING

Now that we have our discretized modeling paradigm and factor model representation as generically represented by equations (9.4) and (9.5), we can move to the three step procedure to derive the absence of arbitrage equations. To guide the development, we will use the familiar case of a European call option c written on a stock S. As usual, let's begin by classifying the variables in our model.

Step 1: Classify Variables
Our classification of variables is

$$\begin{array}{ccc} \text{Tradables} & \text{Underlying Variables} & \text{Factors} \\ B, S, c(S) & r_0, S & \Delta t, Z. \end{array}$$

The factor models for S and c follow from equations (9.4) and (9.5) in the previous section, and will be represented generically as

$$\begin{aligned} \Delta S &= \mathcal{A}_S \Delta t + \mathcal{B}_S Z \\ \Delta c &= \mathcal{A}_c \Delta t + \mathcal{B}_c Z. \end{aligned}$$

For the risk-free asset, we will use the discretized model

$$\Delta B = r_0 B \Delta t.$$

(A different "exact" discrete representation of the bond is given in Exercise 9.1.)

Step 2: Construct a Tradable Table
With our classification of variables, we can write the tradable table as

$$\begin{array}{cc} \text{Prices} & \text{Value Change Factor Models} \\ ----\ \mid & ----------- \end{array}$$

$$\begin{bmatrix} B \\ S \\ -- \\ c \end{bmatrix} \qquad \Delta \begin{bmatrix} B \\ S \\ -- \\ c \end{bmatrix} = \begin{bmatrix} r_0 B \\ \mathcal{A}_S \\ -- \\ \mathcal{A}_c \end{bmatrix} \Delta t + \begin{bmatrix} 0 \\ \mathcal{B}_S \\ -- \\ \mathcal{B}_c \end{bmatrix} Z.$$

In this case, we use the discretized version of the value change of the tradables.

Step 3: Apply the Price APT

Finally, applying the Price APT equation to the marketed tradables from above gives

$$\begin{bmatrix} r_0 B \\ \mathcal{A}_S \end{bmatrix} = \begin{bmatrix} B \\ S \end{bmatrix} \lambda_0 + \begin{bmatrix} 0 \\ \mathcal{B}_S \end{bmatrix} \lambda_1,$$

and solving leads to

$$\lambda_0 = r_0, \quad \lambda_1 = \frac{\mathcal{A}_S - \lambda_0 S}{\mathcal{B}_S}. \tag{9.6}$$

As in the continuous case, λ_0 is the risk-free rate or the short rate driving a money market account. On the other hand, the equation for λ_1 indicates how to "calibrate" the market price of risk for the binary factor Z, as explained next.

9.2.1 Single Step Calibration Equation

We can view the formula for λ_1 in (9.6) as a calibration equation that utilizes the marketed tradables that depend on Z. Substituting in for $\mathcal{A}_S$ and $\mathcal{B}_S$ from equation (9.5) leads to the general "single step calibration" formula,

$$\lambda_1 = \frac{1}{\Delta t} \left[\frac{S^u + S^d - 2S(1 + \lambda_0 \Delta t)}{S^u - S^d} \right]. \tag{9.7}$$

This equation indicates that if we know the value of the marketed tradable S at the nodes of the binary model, then it can be used to determine the market price of risk λ_1 via the above equation. Once we have this market price of risk, we can move to the pricing of the derivative.

9.2.2 Single Step Pricing Equation

Once the market prices of risk have been calibrated, we can use them in the Price APT for c, which is

$$\mathcal{A}_c = c\lambda_0 + \mathcal{B}_c \lambda_1.$$

Applying the definition of $\mathcal{A}_c$ and $\mathcal{B}_c$ from equation (9.5) with c in place of S and substituting leads to

$$\frac{1}{2\Delta t} \left(c^u + c^d - 2c \right) = c\lambda_0 + \frac{1}{2} \left(c^u - c^d \right) \lambda_1.$$

To use this equation to price the derivative c, we can solve for c at the beginning of the period in terms of c^u and c^d at the end of the period, giving

$$c = \frac{1}{(1 + \lambda_0 \Delta t)} \left[\left(\frac{c^u + c^d}{2} \right) - \left(\frac{c^u - c^d}{2} \right) \lambda_1 \Delta t \right]. \tag{9.8}$$

One might call this the "single step pricing" form of the absence of arbitrage equation.

A diagram indicating the single step calibration and pricing sides of the absence of arbitrage equation is given in Figure 9.5.

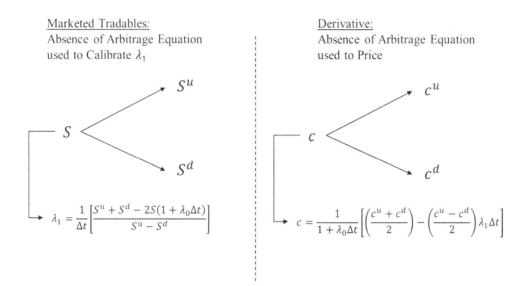

Marketed Tradables:
Absence of Arbitrage Equation
used to Calibrate λ_1

Derivative:
Absence of Arbitrage Equation
used to Price

$$\lambda_1 = \frac{1}{\Delta t}\left[\frac{S^u + S^d - 2S(1 + \lambda_0 \Delta t)}{S^u - S^d}\right]$$

$$c = \frac{1}{1 + \lambda_0 \Delta t}\left[\left(\frac{c^u + c^d}{2}\right) - \left(\frac{c^u - c^d}{2}\right)\lambda_1 \Delta t\right]$$

Figure 9.5 Single step calibration and pricing equations.

9.2.2.1 Interpreting the Pricing Equation

To interpret the single step pricing equation in (9.8), first note that the term $1 + \lambda_0 \Delta t$ represents discounting at the risk-free rate. This is due to the fact that $\lambda_0 = r_0$ is the risk-free rate, which is our reward for the time factor. Since r_0 is annualized, multiplying by Δt just scales it to the proper time increment. Thus, this term discounts at the risk-free rate.

   If we look inside the square brackets of equation (9.8), we see two terms. The first term

$$\left(\frac{c^u + c^d}{2}\right)$$

is the simple average of c at the end of the time period Δt. The second term

$$\left(\frac{c^u - c^d}{2}\right)\lambda_1 \Delta t$$

is the discount for risk. This is because

$$\left(\frac{c^u - c^d}{2}\right)$$

essentially measures the effect of the factor on the outcomes of c. That is, the up and down states are due to the uncertainty in the factor Z. When $Z = 1$, we obtain the up value c^u, and when $Z = -1$, we obtain the down value c^d. The spread between those, $c^u - c^d$, is the extent to which the factor Z affects the value of the derivative. Recall that λ_1 is the market price of risk for Z, and thus determines the discount in the price for the risk exposure.

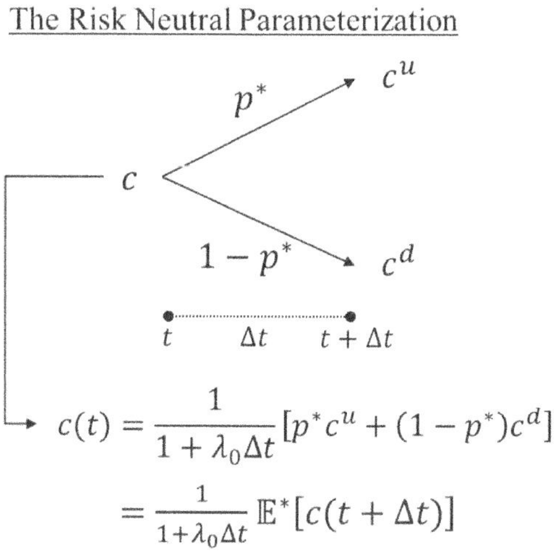

The Risk Neutral Parameterization

$$c(t) = \frac{1}{1 + \lambda_0 \Delta t}[p^* c^u + (1 - p^*)c^d]$$

$$= \frac{1}{1 + \lambda_0 \Delta t}\, \mathbb{E}^*[c(t + \Delta t)]$$

Figure 9.6 Risk neutral parameterization.

9.2.3 Risk Neutral Parameterization

The single step pricing equation can be reparameterized into the so-called *risk neutral* form as

$$c \;=\; \frac{1}{(1 + \lambda_0 \Delta t)}\left[\left(\frac{c^u + c^d}{2}\right) - \left(\frac{c^u - c^d}{2}\right)\lambda_1 \Delta t\right]$$

$$=\; \frac{1}{(1 + \lambda_0 \Delta t)}\left[\frac{(1 - \lambda_1 \Delta t)}{2} c^u + \frac{(1 + \lambda_1 \Delta t)}{2} c^d\right].$$

If we define

$$p^* = \frac{(1 - \lambda_1 \Delta t)}{2}, \quad 1 - p^* = \frac{(1 + \lambda_1 \Delta t)}{2},$$

then these look like "fake" probabilities with p^* the probability of moving to the up state and $1 - p^*$ the probability of the down state. In the language of modern derivative pricing, these are the *risk neutral* probabilities. When we substitute them into the pricing equation, we can act as if we are pricing by a discounted expectation. That is, we can rewrite the pricing formula as

$$c(t) \;=\; \frac{1}{(1 + \lambda_0 \Delta t)}\left[\frac{(1 - \lambda_1 \Delta t)}{2} c^u + \frac{(1 + \lambda_1 \Delta t)}{2} c^d\right]$$

$$=\; \frac{1}{(1 + \lambda_0 \Delta t)}\left[p^* c^u + (1 - p^*)c^d\right]$$

$$=\; \frac{1}{(1 + \lambda_0 \Delta t)}\mathbb{E}^*[c(t + \Delta t)]$$

where $\mathbb{E}^*[\cdot]$ is the expectation under the risk neutral probabilities. See Figure 9.6 for the binary representation of this parameterization.

This is referred to as the risk neutral pricing formula since it seems to ignore the risk in c, compute its expected value at the end of the period, and discount it as if it were risk free. When someone treats an expected payoff as if it were risk free (by discounting it at the risk-free rate), they are said to be risk neutral.

Now, we know that we are not actually being risk neutral here because p^* and $1 - p^*$ are not the real up and down probabilities. But, writing the pricing equation this way can be very convenient, and this concept will be explored in much further detail in Chapter 10. For now, let us note that finding the risk neutral up probability p^* is equivalent to finding the market price of risk λ_1. That is, we can translate back and forth between them using the relationship

$$p^* = \frac{(1 - \lambda_1 \Delta t)}{2}, \quad \lambda_1 = \frac{1}{\Delta t}(1 - 2p^*).$$

With this in mind, we could also calibrate using the risk neutral probability p^* through the pricing equation,

$$p^* = \frac{(1 + \lambda_0 \Delta t)c - c^d}{c^u - c^d},$$

and then convert it to the market price of risk λ_1. As we move forward in this chapter, I will calibrate using the market price of risk equation for λ_1, but I will also often show the corresponding risk neutral probability p^* just to remind you that they are equivalent.

9.3 FINAL STEP: PRICING ON MULTISTEP BINOMIAL TREES

In the previous sections we detailed a binary discretization of the factor models and the absence of arbitrage equation. The final step is to use this to price a derivative with a specified payoff characteristic.

The discretized factor models describe movements over a single Δt period of time. However, to price a derivative, we often have to specify payoff characteristics at expiration time T that is multiple periods of time from the current time. What this means is that specifying the payoff and then pricing a derivative involves multiple binary steps.

Multiple steps of a binary model create what we call a multiperiod binomial tree, as shown in Figure 9.7. To price we begin by creating a multiperiod binomial factor tree which simply extends the single period binary model over many time periods. In each period, the factor Z is either 1, causing a move to the up state u, or -1, causing a move to the down state d.

Just as in the single period setting, since the underlying variables are functions of the factors, the factor tree drives a binomial tree model for the underlying variables. Finally, the tradables are functions of the underlying variables, and thus the underlying variable tree induces a tree for the tradables. This is shown for the example of a stock S and a call option c in Figure 9.8.

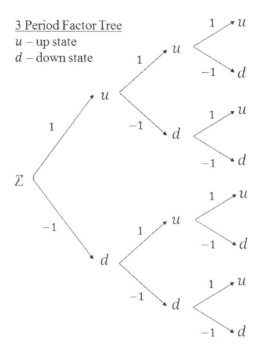

Figure 9.7 Three period binomial factor tree.

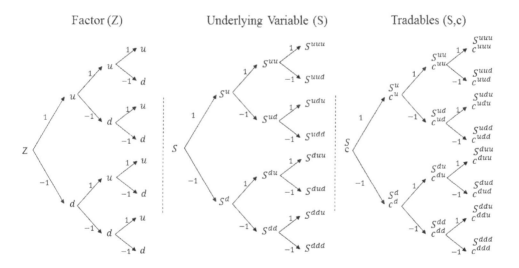

Figure 9.8 Factor, underlying variable, and tradable trees.

Once we have the trees for the tradables, we use the known information about the marketed tradables to determine the market price of risk λ_1 at each node in the tree. For example, the single step calibration equation (9.7) can be used to determine λ_1. This is the calibration phase.

Once the market price of risk has been determined, we move to the tree for the derivative, and begin by applying the known payoff characteristics. We then work, often backward since the payoff is usually specified at the final expiration time T, through the tree applying the single step pricing equation (9.8) until the initial price of the derivative is reached. This process is depicted in Figure 9.9.

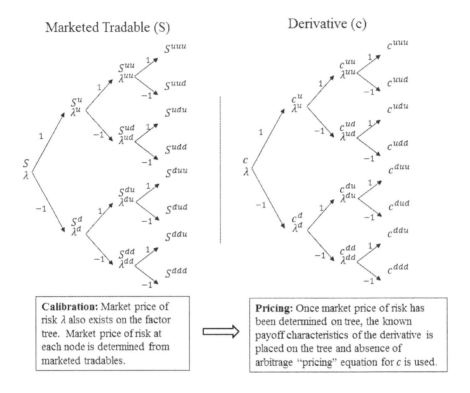

Figure 9.9 Calibration and pricing on the factor tree.

Before proceeding with some examples to illustrate this procedure, we discuss one important simplification that sometimes allows us to reduce the computation required.

9.3.1 Collapsing Trees to Recombining Lattices

A lattice is a recombining tree. In some cases, the tree for the underlying variables and tradables can be made to recombine, and this leads to quite dramatic computational savings when there are many steps in the tree. To illustrate this idea, we consider a call option on a stock that follows geometric Brownian motion.

Geometric Brownian motion with constant coefficients for μ and σ discretizes so that multiple steps move along a recombining lattice. To see this, consider the fact that both models of geometric Brownian motion from Section 9.1.2 (the discretized factor model and the exact solution discretization) are of the form

$$S^u = uS, \quad S^d = dS.$$

In the first case, we have

$$u = \mu \Delta t + \sigma \sqrt{\Delta t}, \quad d = \mu \Delta t - \sigma \sqrt{\Delta t} \tag{9.9}$$

and in the second case, we have

$$u = e^{\left((\mu - \frac{1}{2}\sigma^2)\Delta t + \sigma \sqrt{\Delta t}\right)}, \quad d = e^{\left((\mu - \frac{1}{2}\sigma^2)\Delta t - \sigma \sqrt{\Delta t}\right)}. \tag{9.10}$$

Models of this form recombine as depicted in Figure 9.10, which shows the lattice for the stock over two time periods. Recombination of the nodes occurs at $S^{ud} = S^{du}$ which corresponds to either the stock first moving up and then down or the stock first moving down and then up. In either case, the value of the stock is $S^{ud} = S^{du} = udS$. When this recombination occurs, it leads to the modeling picture as shown

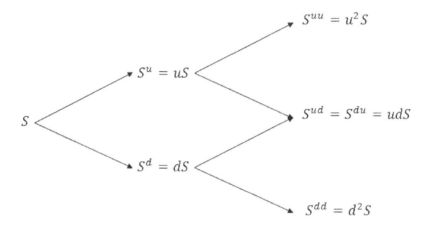

Figure 9.10 Recombining stock lattice.

in Figure 9.11. In this case, all calibration and pricing calculations can take place on the *lattice* for the tradables. It is advantageous to exploit this lattice structure when possible.

To illustrate the ideas presented so far, in the sections that follow we will consider three numerical examples of increasing complexity. The first involves the pricing of call and put options on a stock that follows geometric Brownian motion, where we will be able to reduce computation by using a binomial lattice as just explained. The next two examples price interest rate derivatives using binomial tree structures. In the first we model a CIR driven term structure, while the second considers a single factor LIBOR Market Model.

9.4   PRICING OPTIONS ON A STOCK LATTICE

In this section, we apply the framework given in the previous sections to demonstrate the pricing of options on a stock following geometric Brownian motion.

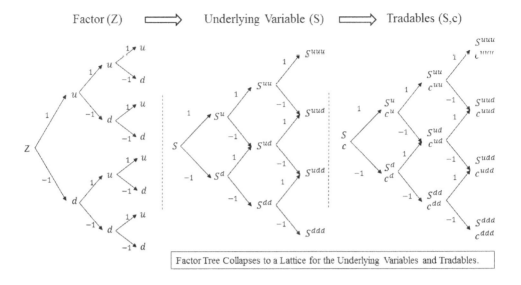

Figure 9.11 Factor tree reduces to lattice for underlying variables and tradables.

9.4.1 Modeling the Stock

Consider a non-dividend paying stock S that follows geometric Brownian motion,

$$dS = \mu S dt + \sigma S dz,$$

with parameter values $\mu = 0.1$ and $\sigma = 0.3$ and current price $S = \$10$. The risk-free rate is taken to be $r_0 = 0.05$.

Consider a binary model discretization of the stock as given in Section 9.1.2.2, equation (9.3), using a time step of $\Delta t = 0.06$. This corresponds to

$$S^u = uS, \quad S^d = dS,$$

where

$$u = e^{\left(\left(\mu - \frac{1}{2}\sigma^2\right)\Delta t + \sigma\sqrt{\Delta t}\right)} = 1.07980956, \qquad (9.11)$$

$$d = e^{\left(\left(\mu - \frac{1}{2}\sigma^2\right)\Delta t - \sigma\sqrt{\Delta t}\right)} = 0.93222163. \qquad (9.12)$$

Assuming an expiration time of $T = 0.3$ and starting from the initial price $S(0) = \$10$, we can build a recombining binomial lattice with discrete times $i = 0 \ldots 5$ corresponding to increments of length $\Delta t = 0.06$ as shown in Figure 9.12. In the spreadsheet-style layout of Figure 9.12, moving straight across from right to left represents an up move ($S \to uS$), and moving diagonally down and to the right represents a down move ($S \to dS$).

Stock Lattice

0	1	2	3	4	5
10.000	10.798	11.660	12.590	13.595	14.680
	9.322	10.066	10.870	11.737	12.674
		8.690	9.384	10.133	10.942
			8.101	8.748	9.446
				7.552	8.155
					7.040

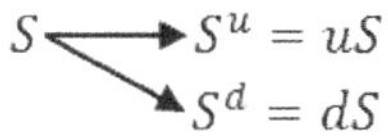

S $S^u = uS$

$S^d = dS$

Figure 9.12   Stock lattice.

9.4.2   Calibration Using Marketed Tradables

As marketed tradables, we have the stock lattice shown in Figure 9.12, and a risk-free asset with risk-free rate 0.05 that follows the discretized model

$$\Delta B = r_0 B \Delta t.$$

We must use these two marketed tradables to calculate the market prices of risk trees (or lattices in this case) for λ_0 and λ_1. However, due to the structure in this specific problem, the values of λ_0 and λ_1 are the same at every node on the lattice. That is, we only need to compute a single value for λ_0 and λ_1 that applies across the entire lattice.

From the results of our discretized Price APT equations (9.6), we know that at every node,

$$\lambda_0 = r_0 = 0.05.$$

Moreover, at every node, we also need to calibrate λ_1 using the single step calibration equation,

$$\lambda_1 = \frac{1}{\Delta t} \left[\frac{S^u + S^d - 2S(1 + \lambda_0 \Delta t)}{S^u - S^d} \right].$$

In this case, using the fact that $S^d = dS$ and $S^u = uS$ provides great simplification. The market price of risk λ_1 is independent of the current price level S, and the formula for λ_1 at every node reduces to

$$\lambda_1 = \frac{1}{\Delta t} \frac{(u + d - 2(1 + \lambda_0 \Delta t))}{(u - d)}. \tag{9.13}$$

Substituting in the up and down values, u and d from equations (9.11) and (9.12), as well as $\lambda_0 = 0.05$, and evaluating gives

$$\lambda_1 = 0.040865.$$

Thus, we have calibrated the lattice by finding the market prices of risk. Finally, if desired, we can also solve for the risk neutral up probability p^* as

$$p^* = \frac{1 - \lambda_1 \Delta t}{2} = \frac{1 - (0.040865)(0.06)}{2} = 0.480.$$

9.4.3 Pricing a European Call Using the Calibrated Lattice

Now that we have the market prices of risk, we can use them to price. Consider a European call option on the stock with expiration $T = 0.3$ and strike $K = \$10$. To price this derivative, we need to use the absence of arbitrage pricing equation on the lattice. Recall that the single step absence of arbitrage pricing formula was given in equation (9.8) as

$$c = \frac{1}{(1 + \lambda_0 \Delta t)} \left[\left(\frac{c^u + c^d}{2} \right) - \left(\frac{c^u - c^d}{2} \right) \lambda_1 \Delta t \right]. \tag{9.14}$$

This equation computes the value of the option at the beginning of a time step given that we know the values at the end of the time step (refer back to Figure 9.5). This is where the payoff characteristics of the specific option come into play.

At expiration $T = 0.3$, which is the fifth time step ($i = 5$) in our lattice, the payoff of the call option is

$$c = \max\{S - 10, 0\}.$$

This can be computed easily for all nodes of the stock lattice at time step $i = 5$, as shown in the last column (in bold) in Figure 9.13. For example, the top entry in the fifth time step column of Figure 9.13 was calculated as

$$c = \max\{S - 10, 0\} = \max\{14.680 - 10, 0\} = 4.680.$$

European Call Option				Strike:	10.000
0	1	2	3	4	5
0.757	1.199	1.827	2.650	3.625	4.680
	0.353	0.627	1.078	1.767	2.674
		0.103	0.215	0.450	0.942
			0.000	0.000	0.000
				0.000	0.000
					0.000

Figure 9.13 European call lattice.

Once the last column is determined from the payoff characteristics of the option, in this case a European call option, we can move backward one column at a time through the lattice using the single step pricing equation (9.14). For example, to compute the value of the call option at the top of the column corresponding to the fourth time step ($i = 4$), we evaluate

$$
\begin{aligned}
c &= \frac{1}{(1 + \lambda_0 \Delta t)} \left[\left(\frac{c^u + c^d}{2} \right) - \left(\frac{c^u - c^d}{2} \right) \lambda_1 \Delta t \right] \\
&= \frac{1}{(1 + 0.05(0.06))} \left[\left(\frac{4.680 + 2.674}{2} \right) - \left(\frac{4.680 - 2.674}{2} \right) (0.040865)(0.06) \right] \\
&= 3.625.
\end{aligned}
$$

By working backward in this manner, one column at a time, we eventually reach the beginning of the call lattice to obtain the current absence of arbitrage value of the call option, $c = 0.757$.

9.4.4 Pricing American Options

Once the lattice for the underlying variable, in this case the stock S, has been constructed and we have completed the calibration phase using the marketed tradables (in this case using just the stock S), the lattice can be used to price many derivatives just by changing the payoff and recomputing on the lattice. In this section, we show how to use our lattice to price American options.

An American call gives the holder the right to buy a stock for the strike price K on *or before* the expiration time T. That is, the difference between an American call option and a European call option is that the holder of the American can exercise at *any time* prior to and including expiration, whereas the holder of the European can exercise only at expiration. An American put option is defined similarly, allowing the holder to exercise and sell at the strike price K at any time up to and including expiration.

The ability to exercise early (i.e., prior to expiration) can add value to an American option over its European counterpart. Moreover, this changes the payoff characteristic of the option. In particular, the payoff of an American option depends on when the holder decides to exercise. In general, we value the American option under the assumption that the holder exercises it to maximize value.

This is easily accommodated with pricing on a binomial lattice. As we work our way backward through the lattice, we simply compare the value of the option with its early exercise value. If the early exercise value is greater, we replace the value of the option with its early exercise value at that node and continue with the binomial lattice pricing. This procedure is illustrated using the calibrated stock lattice from the previous example to price an American put with strike $K = 10$. Its European counterpart is also shown for comparison.

The results are shown in Figure 9.14. The difference between the European put and the American put is that with the European, the payoff is $\max\{K - S, 0\}$ only at time T (time step $i = 5$), whereas with the American put, if the option holder so decides, the option can be exercised at any time for $\max\{K - S, 0\}$. This is easily incorporated into the American put lattice by starting at the final column. If you have reached the final column and still hold the option, it will be worth $\max\{K - S, 0\}$. Thus, the final column ($i = 5$) is the same for both the American and European put options.

Now, step back to column $i = 4$. First calculate the value of the option based on the single step absence of arbitrage pricing equation (9.8). This will be the value if you do not exercise the American put. On the other hand, if you exercise it will be worth $\max\{K - S, 0\}$. Thus, it is rational to choose the larger of these two values. In this manner, at each node in the lattice the value of the option will be the larger

Stock Lattice					
0	1	2	3	4	5
10.000	10.798	11.660	12.590	13.595	14.680
	9.322	10.066	10.870	11.737	12.674
		8.690	9.384	10.133	10.942
			8.101	8.748	9.446
lambda0	0.050			7.552	8.155
lambda1	0.041			7.040	

European Put Option			Strike:		10.000
0	1	2	3	4	5
0.608	0.282	0.077	0.000	0.000	0.000
	0.912	0.472	0.149	0.000	0.000
		1.323	0.772	0.287	0.000
			1.839	1.222	0.554
				2.418	1.845
					2.960

American Put Option			Strike		10.000
0	1	2	3	4	5
0.622	0.286	0.077	0.000	0.000	0.000
	0.936	0.480	0.149	0.000	0.000
		1.361	0.787	0.287	0.000
			1.899	1.252	0.554
				2.448	1.845
					2.960

Figure 9.14 Pricing of European and American puts on stock lattice.

of its early exercise value $\max\{K - S, 0\}$, or continuing to let the option "live" by not exercising it.

For example, referring to the American put in Figure 9.14 at column $i = 4$ in the third row, the value of the option from the single step absence of arbitrage pricing equation can be computed as $p = 0.287$. On the other hand, if the American put option is exercised, it will be worth $p = \max\{10 - 10.133, 0\} = 0$, where the value 10.133 came from the corresponding node in the stock lattice. In this case, it is clearly best to not exercise the option. Hence, for the American put, the value at that node is $p = 0.287$.

However, in the same column $(i = 4)$ at the fourth row, the absence of arbitrage equation gives $p = 1.222$ whereas the early exercise value is $p = \max\{10 - 8.748, 0\} = 1.252$. In this case, it is better to exercise the option than to keep it alive. Thus, the value at this node for the American put is $p = 1.252$.

This procedure is continued backward, at each node using the single step pricing equation (9.8) and comparing it to the early exercise value of the option until the initial value of the option is obtained. From Figure 9.14, we see that the value of the American put is $p = 0.622$ while the value of its European put counterpart is $p = 0.608$.

9.5 PRICING ON A CIR SHORT RATE TREE

In the case of pricing a stock option under geometric Brownian motion in the previous section, many simplifications occurred. This second example involving a single factor CIR short rate model will require us to work through and understand each step in binomial tree calibration and pricing in more detail.

9.5.1 Modeling and Classification of Discretized Variables

We will consider the CIR term structure model encountered in Chapter 7, Section 7.3.2.2, which is a single factor interest rate model based on the following dynamics for the short rate:

$$dr_0 = \kappa(\theta - r_0)dt + b\sqrt{r_0}dz.$$

To be able to compute solutions, we will discretize the short rate dynamics by replacing dt with Δt and dz with $Z\sqrt{\Delta t}$, leading to

$$\Delta r_0 = \kappa(\theta - r_0)\Delta t + b\sqrt{r_0}\sqrt{\Delta t}Z \tag{9.15}$$

We refer to Δt as the time step and Z as the discretized binomial factor. In this example, we will utilize a three step binomial tree. Specifically, our model will extend to a final time of $T = 1$ year and involve 3 time steps of length $\Delta t = 1/3$. For notational convenience, we will refer to variables at various time steps using the index i instead of the time in years t. That is, $r_0(2)$ will be used to denote the value of the short rate at the end of the second time step corresponding to $t = 2\Delta t = 2/3$ years.

For our numerical calculations, we will use the parameter values,

$$\kappa = 0.25, \ \theta = 0.07, \ r_0(0) = 0.05, \ \ b = 0.09, \ \ \Delta t = 1/3.$$

To model this three period market, we begin with the binomial tree for the factor Z, which induces the tree for the short rate from equation (9.15) above. These binomial trees are represented graphically using a spreadsheet-style grid as shown in Figure 9.15. The two binomial trees in the top of the figure show the arrangement of variables through the tree, with the upper left tree showing the factor Z and the upper right tree showing the induced tree for our underlying variable short rate r_0. Note that the superscript of r_0 indicates the path taken through the tree. For example, r_0^{udu} represents the *"udu"* path through the tree that moves up, then down, then up. We will use this convention (without the arrows) for other variables represented on this same tree structure as well.

The tree in the bottom right of Figure 9.15 displays the numerical values of r_0 corresponding to the parameter values above. Again, we will use this convention for the placement of the numerical values to represent quantities on the tree.

In this discretized model, we assume that there are zero-coupon bonds that mature at each time step, $B(i|n)$, $n = 1, 2, 3$, and a money market account B_0. Thus, the variables that make up the model are:

Tradables	Underlying Variables	Factors
$B_0, B(i|1), B(i|2), B(i|3)$	r_0	$\Delta t, Z.$

At this point, we assume that all our bonds and the money market account are marketed tradables.

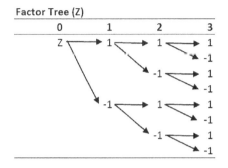

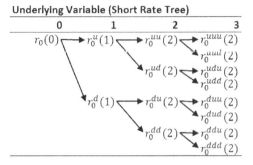

Underlying Variable (Short Rate Tree)			
0	1	2	3
0.0500	0.0633	0.0769	0.0908
			0.0619
		0.0508	0.0641
			0.0407
	0.0400	0.0529	0.0663
			0.0424
		0.0321	0.0446
			0.0260

Figure 9.15 Factor and short rate tree for CIR example displayed in spreadsheet.

9.5.2 Representation of Marketed Tradables

The marketed tradables are zero-coupon bonds of varying maturity. In fact, it is often the case that the price of these bonds is extracted from the term structure of interest rates. As a reminder of how this works, assume that a term structure of interest rates is given using a *simple interest* quoting convention. Let $R(0|n)$ denote the annualized spot rate of interest corresponding to maturity at time step n. Under this convention, the value of a zero-coupon bond that matures at time step n and pays \$1 is given by

$$B(0|n) = \frac{1}{1 + R(0|n)(n\Delta t)}.$$

We assume the current spot rate curve is given in the table below (recall that each time step corresponds to $\Delta t = 1/3$), where we have also calculated the price of the corresponding zero-coupon bonds via the above equation.

Maturity Time:	n	1	2	3
Spot Rate: $R(0|n)$		0.050	0.052	0.054
Zero-Coupon Bond Price: $B(0|n)$		0.9836	0.9665	0.9488

Additionally, assume there exists a money market account $B_0(i)$ that earns the short rate of interest $r_0(i)$ over every time step and follows the dynamics:

$$\Delta B_0 = r_0(i)B_0(i)\Delta t, \tag{9.16}$$

Marketed Tradables

Money Market			
$B_0(0)$	$B_0(1)$	$B_0(2)$	$B_0(3)$
0	1	2	3
1	1.0167	1.0381	1.0647
			1.0647
		1.0381	1.0557
			1.0557
	1.0167	1.0302	1.0484
			1.0484
		1.0302	1.0413
			1.0413

Bond 1			
$B(0|1)$	$B(1|1)$		
0	1	2	3
0.9836	1.0000		
	1.0000		

Bond 2			
$B(0|2)$	$B(1|2)$	$B(2|2)$	
0	1	2	3
0.9665	??	1.0000	
		1.0000	
	??	1.0000	
		1.0000	

Bond 3			
$B(0|3)$	$B(1|3)$	$B(2|3)$	$B(3|3)$
0	1	2	3
0.9488	??	??	1.0000
			1.0000
		??	1.0000
			1.0000
	??	??	1.0000
			1.0000
		??	1.0000
			1.0000

Figure 9.16 Marketed tradables in CIR example.

where the values for $r_0(i)$ are given in the short rate tree of Figure 9.15.

Our final modeling step is to represent the tradables on trees. Figure 9.16 shows all the information we currently have about the tradables placed on binomial trees. To represent the money market account, we use equation (9.16) with an initial value of $B_0(0) = \$1$. Iterating this equation forward using the information provided on the short rate tree leads to the binomial tree for the money market account given in the upper left of Figure 9.16.

On the other hand, for the zero-coupon bonds $B(i|n)$, the only information we have is their initial price $B(0|n)$ given from the term structure of interest rates, and the fact that at maturity they pay a single dollar, $B(n|n) = 1$. This is displayed on binomial trees for the zero-coupon bonds of maturity $n = 1$, 2, and 3 in the upper right and lower trees of Figure 9.16, where "??" is used to indicate unknown values. Since these are the trees for the marketed tradables, this is the information we must use to calibrate the market prices of risk.

9.5.3 Calibration Using Marketed Tradables

Roughly speaking, we can think of the binomial trees for our marketed tradables in Figure 9.16 as replacements for the elements of a tradable table. But, in this case we can't apply the single step calibration equation (9.7) to determine the market price of risk $\lambda_1(i)$ at tree nodes because of all the missing values on the zero-coupon bond trees. So, instead of single step calibration, we have to resort to the absence of arbitrage pricing equation applied across all of the zero-coupon bond trees to calibrate. This is done as follows.

Applying the Price APT to each step of the binomial tree for our tradables results in the single step absence of arbitrage pricing equation (9.8). Imposing this on the zero-coupon bonds over a single time step gives the following equation at each node:

$$B(i|n) = \frac{1}{1 + \lambda_0(i)\Delta t}\left(\frac{B^u(i+1|n) + B^d(i+1|n)}{2} - \frac{B^u(i+1|n) - B^d(i+1|n)}{2}\lambda_1(i)\Delta t\right).$$

The calibration step uses this equation in conjunction with the information in the marketed tradable trees of Figure 9.16 to determine the values of the market prices of risk.

More specifically, note that the market prices of risk λ_0 and λ_1 exist on the factor induced binomial tree as shown in Figure 9.17. In this specific example, the information in the marketed tradables will only allow us to determine λ_0 up to time step 2 and λ_1 up to time step 1, as indicated.

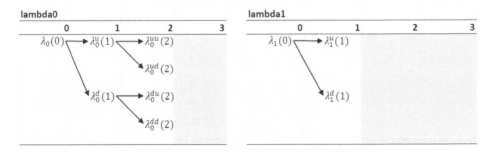

Figure 9.17 Trees for market prices of risk in CIR example.

One (not very good) method to calibrate the market price of risk trees in Figure 9.17 to the marketed tradables in Figure 9.16 is to initially select random values for the market prices of risk on their trees. Next, we could apply the absence of arbitrage equation (9.8) at each node of the trees for the marketed tradables, and finally attempt to change the values of the market prices of risk on their trees until we obtain the correct prices of the marketed tradables on their respective trees. This could be accomplished via an optimization problem. However, it turns out that there is a much easier and faster method to calibrate the market prices of risk to the marketed tradable trees. This is described next.

9.5.4   Market Price of Risk Calibration Algorithm

The following algorithm provides a fast method to calibrate the market prices of risk λ_0 and λ_1 so that all zero-coupon bonds (marketed tradables) are correctly priced. The basic idea of the algorithm is that first the market price of time tree for λ_0 is assigned by noting that it is equal to the short rate tree for r_0. This follows from the basic absence of arbitrage relationships in equation (9.6).

To calibrate the tree for λ_1, an assumption is made that $\lambda_1(i)$ is only a function of time step i and thus is the same value for each node on the tree at time i. With this assumption, by starting with $n = 2$ and increasing n at each loop, the algorithm varies the market price of risk at time $i = n - 2$ (i.e., $\lambda_1(n-2)$) until the zero-coupon bond of maturity n (i.e., $B(0|n)$) is correctly priced.

Calibration Procedure:

Initialization:
The algorithm assumes that the market prices of the zero-coupon bonds $B(0|n)$ for $n = 1 \ldots N$ are known and specified as $B^m(n)$.

- Set the market price of time tree $\lambda_0(i)$ equal to the generated short rate tree $r_0(i)$ for $i = 0 \ldots N - 1$.

- Set $\lambda_1(i) = 0$ for $i = 0 \ldots N - 2$.

Calibration Loop:

For $n = 2 \ldots N$

- Fix all the values of $\lambda_1(i)$ for $i \neq n - 2$. Apply the absence of arbitrage pricing formula to each node of the binomial tree for $B(i|n)$ and compute the initial price $B(0|n)$. Vary $\lambda_1(n - 2)$ until the initial value of $B(0|n)$ on the tree is equal to its market price of $B^m(n)$.

- Set $p^*(n - 2) = \frac{1 - \lambda_1(n-2)\Delta t}{2}$.

Loop n

Note that in the above calibration procedure, we use the bond that matures at time n to determine the market price of risk $\lambda_1(n - 2)$ at time $i = n - 2$. This is because the last step in the tree for a zero-coupon bond leads to a certain payoff of 1 in either the up or down state. Thus, there is no uncertainty in the last step and it cannot provide any information about the market price of risk $\lambda_1(n-1)$ for the factor Z at that time. Thus, we must move back two time steps from the maturity time of the bond to determine information about the market price of risk at time $i = n - 2$.

Applying this calibration procedure to our CIR term structure example leads to the calibrated market price of risk trees at the bottom of Figure 9.18. Note that the market price of risk λ_1 is negative, as we would expect, since an increase in the short rate leads to a decrease in the price of a zero-coupon bond. Moreover, since the procedure determines the market prices of risk, it also allows us to completely fill out the trees for our zero-coupon bonds using the single step pricing formula. The completed trees appear in the top portion of Figure 9.18.

(As an aside, at this point you might want to contrast the assumption made here regarding $\lambda_1(i)$ taking the same value for all nodes at time i with our approach to the continuous time CIR model from Chapter 7, Section 7.3.2.2. There, we made a different assumption about the market price of risk λ_1, and represented it as proportional to $\sqrt{r_0}$. See equation (7.31).)

9.5.5 Pricing Derivatives on CIR Calibrated Trees

Finally, once we have the market prices of risk, we can use them to price any other derivative on the tree via the absence of arbitrage single step pricing equation (9.8). In this specific example, because we only have the market price of risk λ_1 up to time step 1, we are only able to price random payoffs up to time step 2.

To demonstrate the pricing of a derivative, consider a caplet and a floorlet on the short rate r_0 with a strike rate $K = 0.05$ that expires at time $i = 2$. Assume that the principal is $P = \$1,000,000$ and the natural time lag is $\tau = \Delta t = 1/3$. Refer to Chapter 7, Section 7.5.4.

To price this caplet and floorlet, let's recall their payoffs and the timing of the interest payment to which they correspond. Since expiration is at time step $i = 2$, the interest rate for the next payment is set at time step $i = 2$ and the interest payment in an amount of $P\tau r_0(2)$ will occur at time step $i = 2$ plus the natural time lag τ. Since τ is equal to the time step of $\Delta t = 1/3$, this interest payment will occur at time step $i = 3$.

Now, the caplet will allow us to reduce this interest payment by an amount of $P\tau \max\{r_0(2) - K, 0\}$. Thus, the payoff of the caplet occurs at the same time as the interest payment, $i = 3$, and is

$$c(3) = P\tau \max\{r_0(2) - K, 0\}.$$

In a similar manner, the payoff of a floorlet is

$$f(3) = P\tau \max\{K - r_0(2), 0\}$$

since it places a floor on the interest rate for the interest payment.

Finally, since both of these payoffs are known at time $i = 2$ but occur at time $i = 3$, we can discount them back to time $i = 2$ to create an equivalent payoff value at $i = 2$ of

$$c(2) = \frac{P\tau}{1 + r_0(2)\Delta t} \max\{r_0(2) - K, 0\}, \quad f(2) = \frac{P\tau}{1 + r_0(2)\Delta t} \max\{K - r_0(2), 0\}.$$

To price the caplet and floorlet, we specify these payoff values on the binomial

Marketed Tradables

Money Market

$B_0(0)$	$B_0(1)$	$B_0(2)$	$B_0(3)$
0	1	2	3
1	1.0167	1.0381	1.0647
			1.0647
		1.0381	1.0557
			1.0557
	1.0167	1.0302	1.0484
			1.0484
		1.0302	1.0413
			1.0413

Bond 1

$B(0|1)$	$B(1|1)$		
0	1	2	3
0.9836	1.0000		
	1.0000		

Bond 2

$B(0|2)$	$B(0|2)$	$B(0|2)$	$B(0|2)$
0	1	2	3
0.9665	0.9793	1.0000	
		1.0000	
	0.9868	1.0000	
		1.0000	

Bond 3

$B(0|3)$	$B(1|3)$	$B(2|3)$	$B(3|3)$
0	1	2	3
0.9488	0.9584	0.9750	1.0000
			1.0000
		0.9834	1.0000
			1.0000
	0.9726	0.9827	1.0000
			1.0000
		0.9894	1.0000
			1.0000

Market Prices of Risk

lambda0

$\lambda_0(0)$	$\lambda_0(1)$	$\lambda_0(2)$	
0	1	2	3
0.0500	0.0633	0.0769	
		0.0508	
	0.0400	0.0529	
		0.0321	

lambda1

$\lambda_1(0)$	$\lambda_1(1)$		
0	1	2	3
-0.3851	-0.4080		
	-0.4080		

Risk Neutral "up" Probability p*

$p^*(0)$	$p^*(1)$		
0	1	2	3
0.5642	0.5680		
	0.5680		

Figure 9.18 Calibrated bonds and market price of risk trees in CIR example.

tree at time $i = 2$, as provided in the caplet and floorlet trees at the bottom of Figure 9.19, based on the value of the short rate tree at time $i = 2$, which is provided in the upper left tree of Figure 9.19. To provide a specific example of this payoff calculation, using the top value in the short rate tree at time step $i = 2$, $r_0(2) = 0.0769$, the caplet payoff at that time and node using $P = \$1,000,000$, $\tau = 1/3$, and strike rate $K = 0.05$ is

$$c(2) = \frac{(1,000,000)(1/3)}{1 + (0.0769)(1/3)} \max\{0.0769 - 0.05, \ 0\} = 8748.016.$$

Short Rate Tree

0	1	2	3
0.0500	0.0633	0.0769	
		0.0508	
	0.0400	0.0529	
		0.0321	

Caplet	Strike:	0.05	
0	1	2	3
2991.550	4973.486	8748.016	
		253.503	
	540.195	963.740	
		0.000	

Floorlet	Strike:	0.05	
0	1	2	3
1076.084	0.000	0.000	
		0.000	
		2510.306	0.000
		5888.491	

Figure 9.19 Pricing tree for floorlet and caplet in CIR example.

Once the payoffs have been specified, the single step absence of arbitrage pricing equation (9.8), using the calibrated market prices of risk λ_0 and λ_1 at each node in Figure 9.18, is applied step by step backward until the initial node is reached. Referring to the completed trees at the bottom of Figure 9.19, we obtain prices of $c(0) = 2991.550$ for the caplet and $f(0) = 1076.084$ for the floorlet.

9.6 SINGLE FACTOR LIBOR MARKET MODEL

In this final example, we consider a single factor LIBOR Market Model, as in Chapter 7, Section 7.5. Take note that this is a difficult example that should only be embarked upon if you have mastered the previous examples and are looking for a challenge. It is not for the faint of heart!

If you recall, the LIBOR Market Model is an interest rate model in which forward rates are used as the underlying variables. Here, we will consider a model that is driven by a single factor, but we will still have many forward rates to model. An interesting fact about this model is that the underlying variables (forward rates) cannot be modeled independently because they are connected through the absence of arbitrage conditions. This means that we can specify *some* information about the underlying variables, but the calibration procedure that uses the prices of bonds as our marketed tradables will be required to *complete* the specification of the underlying variable forward rate parameters.

9.6.1 LMM Forward Rates and Calibration

We start by discretizing a continuous time model using our time step for discretization Δt as the natural time lag τ in the LMM notation. That is, in continuous time, let $F(t|(n-1)\Delta t, n\Delta t)$ be the forward rate between time $(n-1)\Delta t$ and $n\Delta t$ as seen from the current time t. We will assume that this forward rate follows the dynamics,

$$dF(t|(n-1)\Delta t, n\Delta t) = F(t|(n-1)\Delta t, n\Delta t)\left(a_n(t)dt + b_n(t)dz\right).$$

Note that in this model all of the forward rates are driven by the same single factor dz. Moreover, the forward rate volatility parameter $b_n(t)$ will be allowed to be a function of time t, while the drift parameter $a_n(t)$ will be allowed to be a function of time *and* the current value of the forward rate $F(t|(n-1)\Delta t, n\Delta t)$. This potential dependence of the drift parameter on the forward rate is not shown explicitly in the above equation (just to keep notation simple). It will mean that the discretized version of the drift parameter $a_n(t)$ will live on a binomial tree as well.

To provide motivation for some of the steps that will need to be taken to calibrate this model, recall that according to equation (7.88) from Chapter 7, Section 7.5, a_n and b_n cannot be chosen independently across all the forward rates. That is, the forward rate parameters need to satisfy equation (7.88) with $\rho_{nk} = 1$ (since here we are using a single factor dz across all forward rates instead of many correlated dz_n's), leading to the relationship,

$$a_n - b_n \left(\sum_{k=1}^{n} \frac{\Delta t b_k F(t|(k-1)\Delta t, k\Delta t)}{(1 + \Delta t F(t|(k-1)\Delta t, k\Delta t))} \right) = b_n \lambda_1, \quad n > 1. \tag{9.17}$$

This equation suggests that in the discretized model, once b_n and the market prices of risk are known, they fix the value of a_n. Said another way, we are not free to choose the b_n and a_n parameters in this model independently. Instead, they will (at least partially) be determined during the calibration process. With this said, let's proceed to create a discretized binomial tree model.

9.6.2 Modeling and Classification of Discretized Variables

As we have done previously, we will discretize the factors using a time step Δt and index by the time step i. Moreover, for notational convenience, let's drop explicit use of the time increment Δt and write $F(i|n-1, n) = F(i\Delta t|(n-1)\Delta t, n\Delta t)$. This leads to the discretized dynamics for the forward rates,

$$\Delta F(i|n-1, n) = a_n(i)F(i|n-1, n)\Delta t + b_n(i)F(i|n-1, n)Z\sqrt{\Delta t}. \tag{9.18}$$

Additionally, in this model we will use a money market account $B_0(i)$ and zero-coupon bonds $B(i|n)$ paying a single dollar at maturity n as our marketed tradables. The classification of all variables in the model is given as

Tradables	Underlying Variables	Factors
$B_0, B(i|1), B(i|2), B(i|3)$	$F(i|0,1), F(i|2,3), F(i|3,4)$	$\Delta t, Z.$

9.6.3 Specification of Bond Prices and Forward Rates

We will begin by assuming that there is a term structure of interest rates under a simple interest quoting convention. Let $R(0|n)$ denote the spot rate of interest corresponding to matrity time n. As in the CIR example of the previous section, the value of a zero-coupon bond that matures at time step n is given by

$$B(0|n) = \frac{1}{1 + R(0|n)n\Delta t}.$$

In this case, we are also interested in forward rates that are related to the zero-coupon bond prices through the recursive equation,

$$B(0|n) = \prod_{i=1}^{n} \frac{1}{1 + F(0|i-1,i)\Delta t} = B(0|n-1)\frac{1}{1 + F(0|n-1,n)\Delta t}, \quad (9.19)$$

where we are also using a simple interest quoting convention for the forward rates. We can use this to solve for the initial forward rates as

$$F(0|n-1,n) = \frac{1}{\Delta t}\left(\frac{B(0|n-1)}{B(0|n)} - 1\right). \quad (9.20)$$

Using exactly the same term structure as given in Section 9.5.2 of the CIR example, the table below calculates the initial zero-coupon bond prices as well as the initial forward rates using equation (9.20):

Maturity Time: n	1	2	3
Spot Rate: $R(0|n)$	0.050	0.052	0.054
Zero-Coupon Bond Price: $B(0|n)$	0.9836	0.9665	0.9488
Forward Rate: $F(0|n-1,n)$	0.0500	0.0531	0.0561

Additionally, we assume that there exists a money market account $B_0(i)$ that earns the short rate of interest $r_0(i)$ over every time step. As a reminder, the money market account follows the dynamics

$$\Delta B_0 = r_0(i)B_0(i)\Delta t, \quad (9.21)$$

with $r_0(i) = F(i|i, i+1)$. That is, the short rate is equal to the forward rate one time step before the forward rate's final reference time.

Now, normally we would specify all the parameters governing the forward rates $F(i|n-1,n)$, since they are the underlying variables, and create their binomial trees induced by the binomial factor tree for Z. However, as we emphasized previously, we are not free to specify all the parameters of the forward rates independently. Thus, we will only specify the following forward rate parameters and leave the rest to be determined during the calibration process. We will provide the volatility parameter of each forward rate over time, $b_n(i)$, $i = 0 \ldots n - 2$, and the drift parameter, $a_n(n-2)$ for the forward rate $F(i|n-1,n)$ *only* at time $i = n - 2$, in which case we will assume it applies to *every* node at time $i = n - 2$.

In the case of this numerical example, this corresponds to assigning the following parameter values:

Time Step:	i	0	1
$F(i|1,2)$	$a_2(i)$	−0.04	
	$b_2(i)$	0.35	
$F(i|2,3)$	$a_3(i)$	??	−0.02
	$b_3(i)$	0.38	0.35

Note that the first forward rate $F(0|0,1)$ is just the current spot rate and does not "live" past time step 0. That is, we don't need to specify any parameter values for it. Moreover, each forward rate $F(i|n-1,n)$ only "lives" to time $i = n-1$ and thus only requires parameter values for its dynamics to time $i = n - 2$ to determine its values up to time $i = n - 1$.

For example, for the forward rate $F(i|2,3)$, a complete specification as given above involves the volatility and and drift parameters $b_3(i)$ and $a_3(i)$ for $i = 0$ and 1. Because of the absence of arbitrage constraints, we will specify $b_3(i)$ over its entire life, but $a_n(i)$ will only be directly assigned for $i = 1$. The missing drift value of $a_3(i)$ for $i = 0$ will be determined during the calibration procedure to respect the no-arbitrage condition.

Figure 9.20 displays this information on binomial trees for the underlying variable forward rates. The specification of the volatility of each rate $b_n(i)$ is given in a row above each binomial tree, while the drift values $a_n(i)$ are highlighted below each node since the calibrated values will be allowed to depend on the current state of the binomial tree. Values of the forward rates that can be calculated using the parameters and the dynamics in equation (9.18) are shown on the binomial trees, while missing values that require calibration before they can be computed are indicated with question marks "??".

9.6.4 Calibration Using Marketed Tradables

The zero-coupon bonds $B(i|n)$ and the money market account are our marketed tradables. Initially, we know the payoff values of the zero-coupon bonds, $B(n|n) = 1$ at each node at time $i = n$ and their current prices. We can also specify a portion of the binomial tree for the money market account given the forward rate information that we have. This marketed tradable information is provided in the binomial trees of Figure 9.21.

Factor Tree

Z				
i	0	1	2	3
	Z	1.0000	1.0000	1.0000
				-1.0000
			-1.0000	1.0000
				-1.0000
		-1.0000	1.0000	1.0000
				-1.0000
			-1.0000	1.0000
				-1.0000

Underlying Variables

Forward Rate $F(i|0,1)$

i	0	1	2	3
	0.0500			

Forward Rate $F(i|1,2)$

$b_2(i)$	0.3500			
i	0	1	2	3
$F(i|1,2)$	0.0531	0.0631		
$a_2(i)$	-0.0400			
		0.0417		

Forward Rate $F(i|2,3)$

$b_3(i)$	0.3800	0.3500		
i	0	1	2	3
$F(i|2,3)$	0.0561	??	??	
$a_3(i)$	??	-0.0200		
			??	
		??	??	
		-0.0200		
			??	

Figure 9.20 Factor and underlying variable trees in LMM example.

Marketed Tradables

Money Market

$B_0(0)$	$B_0(1)$	$B_0(2)$	$B_0(3)$
0	1	2	3
1	1.0167	1.0381	??
			??
		1.0381	??
			??
	1.0167	1.0308	??
			??
		1.0308	??
			??

Bond 1 (T = 1/3)

$B(0|1)$	$B(1|1)$		
0	1	2	3
0.9836	1.0000		
	1.0000		

Bond 2 (T = 2/3)

$B(0|2)$	$B(1|2)$	$B(2|2)$	
0	1	2	3
0.9665	??	1.0000	
		1.0000	
	??	1.0000	
		1.0000	

Bond 3 (T = 1)

$B(0|3)$	$B(1|3)$	$B(2|3)$	$B(3|3)$
0	1	2	3
0.9488	??	??	1.0000
			1.0000
		??	1.0000
			1.0000
	??	??	1.0000
			1.0000
		??	1.0000
			1.0000

Figure 9.21 Marketed tradables in LMM example.

The calibration step is to use the information in the marketed tradable trees of Figure 9.21 to determine the values of the market prices of risk on their binomial trees which are depicted in Figure 9.22. That is, we must use the absence of arbitrage

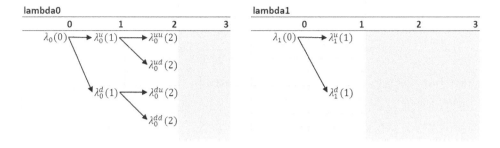

Figure 9.22 Trees for market prices of risk in LMM example.

pricing equation,

$$B(i|n) = \frac{1}{1+\lambda_0(i)\Delta t}\left(\frac{B^u(i+1|n) + B^d(i+1|n)}{2} - \frac{B^u(i+1|n) - B^d(i+1|n)}{2}\lambda_1(i)\Delta t\right)$$

and adjust $\lambda_0(i)$ and $\lambda_1(i)$ on their binomial trees so that the marketed tradables are correctly priced in Figure 9.21. Moreover, at the same time we need to consistently assign the missing forward rate drift parameters $a_n(i)$ in Figure 9.20. Thus, next we show how the forward rate drift is constrained by the absence of arbitrage equation, and also provide a convenient closed form solution that will allow us to calibrate the missing forward rate drift parameters efficiently.

9.6.5   Calibration Equation for the Forward Rate Parameters

To create a calibration equation for the drift parameters $a_n(i)$ of the forward rates, just as in the continuous time case, we will write the absence of arbitrage equation for the tradable bonds in terms of the forward rates in a manner that will "expose" the drift $a_n(i)$. This will allow us to solve for the drift $a_n(i)$ in terms of the other variables.

To start, assume that $i < n - 2$, and that the binomial tree for $B(i|n-1)$ is known to time $i+1$. Moreover, assume that $B(i|n)$, $\lambda_1(i)$, and $\lambda_0(i)$ are also known. Our goal is to compute $a_n(i)$ from this information.

The absence of arbitrage equation at a specific node parameterized by the risk neutral up probability $p^*(i)$ at that node is

$$B(i|n) = \frac{1}{1 + \lambda_0(i)\Delta t}\left(p^* B^u(i + 1|n) + (1 - p^*)B^d(i + 1|n)\right) \tag{9.22}$$

where $p^*(i) = \frac{1-\lambda_1(i)}{2}$.

Now, note that we can recursively write the prices of zero-coupon bonds at the next time step $i + 1$ in the up and down states using a relationship similar to that given in equation (9.19):

$$
\begin{aligned}
B^u(i+1|n) &= B^u(i+1)|n-1)\frac{1}{1+\Gamma''(i+1|n-1,n)\Delta t}, \\
B^d(i+1|n) &= B^d(i+1)|n-1)\frac{1}{1+F^d(i+1|n-1,n)\Delta t}.
\end{aligned}
$$

We also have our discretization of the forward rates in the up and down states as

$$
\begin{aligned}
F^u(i+1|n-1,n) &= F(i|n-1,n)(1+a_n(i)\Delta t+b_n(i)\sqrt{\Delta t}) \\
F^d(i+1|n-1,n) &= F(i|n-1,n)(1+a_n(i)\Delta t-b_n(i)\sqrt{\Delta t}).
\end{aligned}
$$

These equations "expose" the forward rate parameters at each node at time step i. Substituting these into equation (9.22) gives

$$
\begin{aligned}
B(i|n) = \frac{1}{1+\lambda_0(i)\Delta t}\Bigg(&p^*\frac{B^u(i+1)|n-1)}{1+F(i|n-1,n)(1+a_n(i)\Delta t+b_n(i)\sqrt{\Delta t})\Delta t} \\
&+(1-p^*)\frac{B^d(i+1)|n-1)}{1+F(i|n-1,n)(1+a_n(i)\Delta t-b_n(i)\sqrt{\Delta t})\Delta t}\Bigg).
\end{aligned} \quad (9.23)
$$

This equation provides an absence of arbitrage constraint on the parameters of the forward rate dynamics, $a_n(i)$ and $b_n(i)$. In particular, if $a_n(i)$ is the only unknown, we can use this equation to solve for $a_n(i)$ so that no arbitrage will be possible. Roughly speaking, this equation serves the same role as equation (7.88) in the continuous time LMM model.

9.6.5.1 Solving for $a_n(i)$

For calibration purposes, it is convenient to solve for $a_n(i)$ in closed form from equation (9.23) as follows. Define

$$
X = 1 + F(i|n-1,n)(1+a_n(i)\Delta t)\Delta t, \quad Y = b_n(i)(\Delta t)^{3/2}. \quad (9.24)
$$

Next, solve for X in terms of Y by calculating

$$
\hat{b} = \frac{B(i|n-1)}{B(i|n)}, \quad (9.25)
$$

$$
\hat{c} = \frac{Y\left[p^*B^u(i+1|n-1)-(1-p^*)B^d(i+1|n-1)\right]}{B(i|n)(1+\lambda_0(i)\Delta t)} - Y^2, \quad (9.26)
$$

$$
X = \frac{-\hat{b}+\sqrt{\hat{b}^2-4\hat{c}}}{2}. \quad (9.27)
$$

Finally, convert X to the drift parameter $a_n(i)$ via

$$
a_n(i) = \frac{1}{\Delta t}\left(\frac{X-1}{F(i|n-1,n)\Delta t}-1\right). \quad (9.28)
$$

This allows us to calibrate the drift terms of the forward rates, $a_n(i)$, given volatilities $b_n(i)$ and market prices of risk (or equivalently risk neutral probabilities). With this equation in hand, we provide a calibration algorithm for this LMM model next.

9.6.6 Calibration Algorithm for LMM

The following algorithm efficiently calibrates the LMM model to zero-coupon bond prices, and also assigns the missing forward rate parameters.

LMM Calibration Algorithm:

Initialization: The algorithm assumes that zero-coupon bond prices for maturities out to time step N are known: $B(0|n)$ for $n = 1 \ldots N$. It also uses the convention that $B(0|0) = 1$.

- Compute the initial values of the forward rates $F(0|n-1, n)$ for $n = 1 \ldots N$ from equation (9.20).

- The user supplies the following drift and volatility parameter information about the forward rates

$$b_n(i), \; i = 1 \ldots n - 2 \; \text{ and } \; a_n(n-2)$$

for $n = 2 \ldots N$.

Calibration Loop:
For $n = 1 \ldots N$
   For $i = 0 \ldots n - 1$
      For each state in the factor tree at time i, calculate the following:

- When $i < n - 2$, use equations (9.24) through (9.28) to compute $a_n(i)$ at the current state.

- When $i < n - 1$, compute:

$$\begin{cases} F^u(i+1|n-1, n) &= F(i|n-1, n)(1 + a_n(i)\Delta t + b_n(i)\sqrt{\Delta t}) \\ F^d(i+1|n-1, n) &= F(i|n-1, n)(1 + a_n(i)\Delta t - b_n(i)\sqrt{\Delta t}) \\ B^u(i+1|n) &= \frac{B^u(i+1|n-1)}{1+F^u(i+1|n-1,n)\Delta t} \\ B^d(i+1|n) &= \frac{B^d(i+1|n-1)}{1+F^d(i+1|n-1,n)\Delta t}. \end{cases}$$

- If $i = n - 2$, calibrate $\lambda_1(i)$ and $p^*(i)$ by calculating

$$\begin{cases} \lambda_1(i) &= \frac{1}{\Delta t}\left[\frac{B^u(i+1|n)+B^d(i+1|n)-2B(i|n)(1+\lambda_0(i)\Delta t)}{B^u(i+1|n)-B^d(i+1|n)}\right], \\ p^*(i) &= \frac{1-\lambda_1(i)\Delta t}{2}. \end{cases}$$

- If $i = n-1$, set $\lambda_0(i) = F(i|n-1, n)$ and $B^u(i+1|n) = B^d(i+1|n) = 1$.

   Loop i
Loop n

Figures 9.23, 9.24, and 9.25 show the results of the calibration algorithm. In particular, Figure 9.23 shows the completed forward rate trees where the missing values of $a_n(i)$ have now been filled in such that no arbitrage is allowed. Moreover, Figure 9.24 shows the completed binomial trees for all the marketed tradable bonds. Most importantly, Figure 9.25 provides the binomial trees for the calibrated market prices of risk, λ_0 and λ_1, and the risk neutral up probabilities p^*. Once we have the completed market price of risk trees, this enables us to price derivative securities.

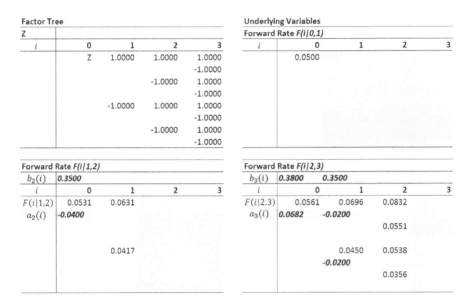

Figure 9.23 Calibrated forward rates in LMM example.

9.6.7 Pricing Derivatives with LMM Calibrated Trees

Once we have the market prices of risk, we can use them to price derivatives using the absence of arbitrage pricing equation (9.8). As in the case of the CIR model, note that because we only have the market price of risk λ_1 up to time step $i = 1$, we are only able to price random payoffs up to time step $i = 2$.

As in the previous CIR example, let's consider a caplet and a floorlet on the short rate r_0 with a strike rate $K = 0.05$ that expires at time $i = 2$. The principal is assumed to be $P = \$1,000,000$ and the natural time lag is $\tau = 1/3$. Exactly as in the CIR example, at time $i = 2$ the payoff of the caplet and floorlet is given by

$$c(2) = \frac{P\tau}{1 + r_0(2)\Delta t} \max\{r_0(2) - K,\, 0\}, \quad f(2) = \frac{P\tau}{1 + r_0(2)\Delta t} \max\{K - r_0(2),\, 0\}.$$

To price, we first compute these payoffs on a binomial tree at time $i = 2$, noting that $r_0(2) = F(2|2,3)$ at each node at time step $i = 2$. Thus, using the values of the underlying forward rate variable $F(2|2,3)$ at time step $i = 2$, we are able to calculate the payoffs of the caplet and floorlet as shown in bold in Figure 9.26.

Once the payoff has been specified, we use the calibrated λ_0 and λ_1 market

Marketed Tradables

Money Market			
$B_0(0)$	$B_0(1)$	$B_0(2)$	$B_0(3)$
0	1	2	3
1	1.0167	1.0381	1.0669
			1.0669
		1.0381	1.0571
			1.0571
	1.0167	1.0308	1.0493
			1.0493
		1.0308	1.0430
			1.0430

Bond 1 (T = 1/3)			
B(0|1)	B(1|1)		
0	1	2	3
0.9836	1.0000		
	1.0000		

Bond 2 (T = 2/3)			
B(0|2)	B(1|2)	B(2|2)	
0	1	2	3
0.9665	0.9794	1.0000	
		1.0000	
	0.9863	1.0000	
		1.0000	

Bond 3 (T = 1)			
B(0|3)	B(1|3)	B(2|3)	B(3|3)
0	1	2	3
0.9488	0.9572	0.9730	1.0000
			1.0000
		0.9820	1.0000
			1.0000
	0.9717	0.9824	1.0000
			1.0000
		0.9883	1.0000
			1.0000

Figure 9.24 Calibrated bond trees in LMM example.

Market Prices of Risk

lambda0			
F(0|0,1)	F(1|1,2)	F(2|2,3)	
0	1	2	3
0.0500	0.0631	0.0832	
		0.0551	
	0.0417	0.0538	
		0.0356	

lambda1			
0	1	2	3
-0.2084	-0.1127		
	-0.1079		

Risk Neutral "up" Probability p*			
0	1	2	3
0.5347	0.5188		
	0.5180		

Figure 9.25 Calibrated market price of risk and risk neutral probability trees.

prices of risk trees in Figure 9.25 and work step by step backward using the single step absence of arbitrage pricing formula (9.8) until the initial node is reached. In

Pricing on the Calibrated Trees

Caplet	Strike		0.05	
0	1	2		3
3587.303	6262.640	10779.075		
		1667.497		
	640.929	1254.531		
		0.000		

Floorlet	Strike:		0.05	
0	1	2		3
2136.377	0.000	0.000		
		0.000		
	4668.336	0.000		
		4733.185		

Figure 9.26 Caplet and floorlet prices using calibrated trees in LMM example.

this case, the completed trees are given in Figure 9.26, and result in the prices of $c(0) = 3587.303$ for the caplet and $f(0) = 2136.377$ for the floorlet.

As you may have gathered from this example, using the binomial tree approach to calibrate and price in the LMM model is not always a simple task. This is a model that in many ways is better suited for the risk neutral pricing methodology, which will be more fully introduced in the next chapter.

9.7 ADDITIONAL COMPUTATIONAL METHODS

To this point, our binomial tree models have only been applied to *single factor* models. What happens when we want to price a multifactor model? Well, the first thing to note is that the framework of this chapter applies equally well to multifactor models as it does to single factor models. For example, two factor and multivariate derivatives have been considered in a lattice framework in [7] and [8]. However, the *amount* of computation increases significantly. In fact, even pricing a single factor model on a binomial tree can be computationally intensive depending on the number of time steps involved. Multifactor models only make that computational burden worse.

There are two other "main" approaches to the computation of solutions, both of which handle multifactor models better than binomial trees. They are the partial differential equation approach and the Monte Carlo approach. In the partial differential equation approach, the continuous time absence of arbitrage equation for a derivative, which usually takes the form of a partial differential equation, is solved numerically using computational methods for partial differential equations. Since the subject of computational methods for solving partial differential equations is quite old and well developed, it provides a reasonable approach for many problems. However, these methods are generally only suitable for models with three or fewer factors. Nevertheless, the argument that the absence of arbitrage equations are partial differential equations and thus we should be using partial differential equation numerical methods is a compelling one. I highly recommend that at some point you supplement your knowledge with some of these methods. Wilmott [53] is a great start.

When the dimension of a problem becomes too large for binomial or partial differential equation methods, one typically resorts to the Monte Carlo approach. This approach is most consistently explained from the risk neutral pricing point of view, which was briefly touched upon in this chapter, and will be introduced in the next. Using the risk neutral parameterization, the pricing equation can be written as an expectation with respect to the risk neutral probabilities. One can then estimate this expectation using Monte Carlo simulations. This approach scales very well with the dimension of the problem, and thus is the most successful method of pricing high-dimensional derivatives. Once again, after you gain a solid understanding of the risk neutral approach (the next chapter provides an introduction), you should make sure to learn about pricing via Monte Carlo simulation from references such as [22, 32].

9.8 SUMMARY

In this chapter, we presented the binomial tree approach to the computation of absence of arbitrage prices. This approach follows naturally from our factor model pricing paradigm upon replacing the random factors with binary approximations. This replacement leads to binomial tree representations for the factors, which in turn induce trees for the underlying variables and tradables. The absence of arbitrage condition then becomes a relationship at each node of the trees and can be used to either calibrate (find the market prices of risk at a node) or price (find the value of the derivative at a node). We demonstrated this methodology for European and American options on a stock following geometric Brownian motion, a single factor CIR short rate term structure model, and a single factor LIBOR Market Model.

I believe that there is great value in emphasizing the tight connection between the binomial tree method and the factor approach, as has been done in this chapter. Working carefully through the examples and then returning to their continuous time counterparts should lead to improved understanding of the models, their assumptions, and their calibration and pricing. Further extensions of binomial tree pricing are found in the exercises, so please don't neglect them!

EXERCISES

9.1 An "Exact" Model for the Risk-Free Bond.

Instead of using the discretized factor model for the bond $\Delta B = r_0 B \Delta t$, we can create an "exact" model using the solution to the continuous bond factor model $dB = r_0 B dt$, which is $B(t) = B(0)e^{r_0 t}$.

(a) Using a discretization of the "exact" bond solution, show that this leads to a discretized version of the form, $\Delta B = \frac{(e^{r_0 \Delta t} - 1)}{\Delta t} B \Delta t$.

(b) Show that this "exact" model leads to $\lambda_0 = \frac{(e^{r_0\Delta t}-1)}{\Delta t}$, and thus in our pricing equations, we would be able to use the replacement $1 + \lambda_0\Delta t = e^{r_0\Delta t}$

9.2 Create a three period binomial lattice with time step $\Delta t = 0.05$ corresponding to the geometric Brownian motion model,

$$dS = 0.2Sdt + 0.4Sdz,$$

with $S(0) = \$20$. Use the binary approximation for geometric Brownian motion in (9.3).

9.3 Solve Exercise 9.2 again, but use the binary approximation given in equation (9.2).

9.4 Price a European call option on a non-dividend paying stock that follows geometric Brownian motion. Assume $N = 2$, $\Delta t = 0.2$, $\mu = 0.1$, $\sigma = 0.2$, and $r_0 = 0.05$. Assume the option has strike price $K = \$10.0$ and the initial value of the stock price is $S(0) = \$10.0$. Proceed in the following steps:

(a) Create the stock lattice using the model in equation (9.3).

(b) Using the stock lattice, compute the market price of risk λ_1 from the stock lattice.

(c) Solve the call lattice.

9.5 Repeat Exercise 9.4 for a European put option.

9.6 Annualizing the Discretized Market Price of Risk.

Show that if λ is the market price of risk corresponding to the discretized binary factor Z, then the market price of risk corresponding to the discretized factor $\Delta z = Z\sqrt{\Delta t}$ will be equal to $\lambda\sqrt{\Delta t}$.

9.7 Simplify the single step market price of risk calibration equation (9.7) when $S^u = uS$ and $S^d = dS$. In particular, show that

$$\lambda_1 = \frac{1}{\Delta t}\left[\frac{u + d - 2(1 + \lambda_0\Delta t)}{u - d}\right].$$

9.8 Calibration using a Futures Price.

Compute the single step calibration equation for λ_1 when the tradable is a futures contract with corresponding underlying variable futures price that at each step moves from f to $f^u = uf$ or $f^d = df$. Specifically, show that

$$\lambda_1 = \frac{1}{\Delta t}\left(\frac{u + d - 2}{u - d}\right).$$

(Hint: Start from the single step pricing equation (9.8) and recall that the price of a futures contract is 0 at the beginning of each period, and that it pays $df = f(t + \Delta t) - f(t)$ at the end of the period.)

9.9 Binomial Model under Continuous Dividends.

Assume that S is a stock that pays continuous dividends at a rate of q and that its price follows geometric Brownian motion with

$$dS = \mu S dt + \sigma S dz.$$

We would like to develop a binary approximation of the form of (9.3). Let u and d be the up and down factors corresponding to S as in (9.10) so that $S^u = uS$ and $S^d = dS$.

(a) Consider the up and down factors, u and d, for the stock price movement given in equation (9.10). Show that the up and down factors for the value of a share v can be written as

$$u' = e^{q\Delta t}u, \quad d' = e^{q\Delta t}d.$$

(b) Use part (a) to derive the single step calibration equation for λ_1 on a stock that pays a continuous dividend. Specifically, show that

$$\lambda_1 = \frac{1}{\Delta t}\left[\frac{u + d - 2e^{-q\Delta t}(1 + \lambda_0\Delta t)}{u - d}\right].$$

9.10 Use the calibrated stock price lattice of Section 9.4, Figure 9.14, to price a European put option with expiration at time step $n = 5$ and strike price $K = \$11$.

9.11 Use the calibrated stock price lattice of Section 9.4, Figure 9.14, to price an American put option with expiration at time step $n = 5$ and strike price $K = \$11$.

9.12 Use the calibrated trees from the CIR example of Section 9.5 to price a caplet. Use a strike rate of $K = 0.04$, principal of $P = \$1,000,000$, natural time lag of $\tau = \Delta t = 1/3$, and expiration at time step $n = 2$.

9.13 Use the calibrated trees from the LMM example of Section 9.6 to price a caplet. Use a strike rate of $K = 0.04$, principal of $P = \$1,000,000$, natural time lag of $\tau = \Delta t = 1/3$, and expiration at time step $n = 2$.

9.14 Computation on a Lattice versus a Tree.

Consider a binomial tree and a recombining binomial lattice with N time steps. To obtain a very rough sense of how the computation involving a tree and a lattice scales, we will calculate and compare the number of nodes at the end of a tree versus a lattice.

(a) For a tree with N time steps, how many nodes exist at the end of the tree (time N) as a function of N?

(b) For a lattice with N time steps, how many nodes exist at the end of the lattice (time N) as a function of N?

(c) Roughly speaking, a calculation needs to be performed at every node of a lattice or tree to compute the price of an option. Let's just compare the number of nodes at the end of the tree or lattice using the results of (a) and (b). Calculate the number of nodes at the end of the tree and lattice in (a) and (b) for $N = 5$, 20, and 100.

9.15 Calibration in LMM.

Show that the formula for $a_n(i)$ in Section 9.6.5.1, equations (9.24) through (9.28), is the solution to equation (9.23). Proceed in the following steps.

(a) Using the definitions

$$X = 1 + F(i|n - 1, n)(1 + a_n(i)\Delta t)\Delta t, \quad Y = b_n(i)\sqrt{\Delta t}\Delta t$$

show that equation (9.23) can be reduced to the quadratic equation

$$X^2 - X\frac{B(i|n-1)}{B(i|n)}$$
$$+Y\left(\frac{p^* B^u(i+1|n-1) - (1-p^*)B^d(i+1|n-1)}{B(i|n)(1+\lambda_0(i)\Delta t)}\right) - Y^2 = 0 \tag{P9.1}$$

where

$$B(i|n - 1) = \frac{1}{1 + \lambda_0(i)\Delta t}\left[p^* B^u(i + 1|n - 1) + (1 - p^*)B^d(i + 1|n - 1)\right].$$

(b) Define $\hat{b} = -\frac{B(i|n-1)}{B(i|n)}$ and

$$\hat{c} = Y\left(\frac{p^* B^u(i + 1|n - 1) - (1 - p^*)B^d(i + 1|n - 1)}{B(i|n)(1 + \lambda_0(i)\Delta t)}\right) - Y^2$$

and write the solution to equation (P9.1) which is quadratic in X in terms of $\hat{b}$ and $\hat{c}$.

(c) Finally, solve for $a_n(i)$ in terms of X to obtain equation (9.28).

CHAPTER 10

The Road to Risk Neutrality

RISK neutral absence of arbitrage pricing is a widely used approach to derivative pricing that, when understood properly, can be extremely powerful. In this chapter, I will explain the basic idea of risk neutral pricing. This introduction does not follow the standard probability-heavy route, but instead motivates risk neutral pricing in a direct manner from the factor approach. I hope you appreciate the simplicity with which we arrive at the concept of risk neutral pricing, and the idea that it can be seen as a natural and logical consequence of the factor approach. In fact, it arises out of a simple question that perhaps may have entered your mind as you worked through this book...

10.1 DO THE FACTORS MATTER?

Here is a provocative question. Do the factors even matter? What I mean by this is that if we look at our derivation of the Price APT, the exact form of the factors doesn't really seem to matter.

To see what I am talking about, let's start from a generic tradable table,

$$
\begin{array}{c|c}
\text{Prices} & \text{Value Change Factor Models} \\
---- & ------------- \\
[\ \mathcal{P}\ ] & d[\ \mathcal{V}\ ] = [\ \mathcal{A}\ ]\,dt + [\ \mathcal{B}\ ]\,dz
\end{array}
$$

and recall that the Price APT absence of arbitrage condition is

$$
\mathcal{A} = \mathcal{P}\lambda_0 + \mathcal{B}\lambda. \tag{10.1}
$$

Thus, absence of arbitrage only depends on the values of $\mathcal{P}$, $\mathcal{A}$, and $\mathcal{B}$, and not on what the exact factors dz are. Therefore, the prices $\mathcal{P}$ are absence of arbitrage for any value change with factor model coefficients $\mathcal{A}$, and $\mathcal{B}$, regardless of what the driving factors actually are! This rather surprising revelation leads us to ponder the following hypothesis.

Hypothesis: *Perhaps by changing the factors from dz to some other random factors $d\psi$, where the $\mathcal{A}$ and $\mathcal{B}$ factor model coefficients of value changes are the same, we can more easily compute the absence of arbitrage prices $\mathcal{P}$.*

To see whether something can be made of this hypothesis, we have to start with a more basic question.

Question: *Given a set of factors dz, what other set of factors $d\psi$ will have the same $\mathcal{A}$ and $\mathcal{B}$ factor model coefficients?*

At first glance, you might be tempted to say that you can choose $d\psi$ to be anything you want. That is, you might think that you could swap out the original dz factors with any other factors $d\psi$ that you want, and it wouldn't affect anything at all because the absence of arbitrage condition "does not see" dz. However, you would soon realize that this is not the case.

Why? Because when we deal with derivative securities, their factor models are determined by an application of Ito's lemma. Thus, the $\mathcal{A}$ and $\mathcal{B}$ representation of price changes does "see" the factors via Ito's lemma. Hence, Ito's lemma puts a constraint on which factors $d\psi$ are consistent with the original factors dz. But it turns out that our basic idea is correct. That is, we can change the factors, up to a point, and it won't affect the absence of arbitrage prices.

Let's be a little more concrete about this this whole idea and start by exploring it in the context of Brownian factors in our standard geometric Brownian motion model.

10.1.1 Brownian Factors

Let $S(t)$ be a stock price under geometric Brownian motion with factor model

$$dS = \mu S dt + \sigma S dz, \tag{10.2}$$

and let $c(S,t)$ be a derivative security. By Ito's lemma we have

$$dc = \left(c_t + \mu S c_S + \frac{1}{2}\sigma^2 S^2 c_{SS} \right) dt + \sigma S c_S dz. \tag{10.3}$$

If we also include a risk-free asset B with risk-free rate r_0, then a tradable table in this situation is

$$
\begin{array}{cc}
\text{Prices} & \text{Value Change Factor Models} \\
\text{-- -- --} \quad | & \text{-- -- -- -- -- -- -- -- -- --} \\
\begin{bmatrix} B \\ S \\ c \end{bmatrix} & d \begin{bmatrix} B \\ S \\ c \end{bmatrix} = \begin{bmatrix} r_0 B \\ \mu S \\ (c_t + \mu S c_S + \frac{1}{2}\sigma^2 S^2 c_{SS}) \end{bmatrix} dt + \begin{bmatrix} 0 \\ \sigma S \\ \sigma S c_S \end{bmatrix} dz.
\end{array}
$$

Let's explore simple methods of changing the factor dz, and determine whether it results in an identical tradable table. We will start by seeing if we can replace the Brownian factor dz with a new Brownian factor plus a drift.

10.1.1.1   Adding a Drift Works

Let's try new factors given by the replacement

$$dz \to d\psi = d\tilde{z} + \beta dt. \tag{10.4}$$

That is, we will replace the original Brownian factor dz by a new factor that is a Brownian $d\tilde{z}$ plus a drift βdt. It is important to consider the new factor to be the entire term $d\psi = d\tilde{z} + \beta dt$ and not just $d\tilde{z}$.

We can ask whether this will lead to a tradable table representation consistent with the tradable table under the original dz. Replacing dz with this factor $d\psi$ creates stock dynamics of the form

$$
\begin{aligned}
dS &= \mu S dt + \sigma S(d\psi) \\
&= \mu S dt + \sigma S(d\tilde{z} + \beta dt) \\
&= (\mu + \sigma \beta S)dt + \sigma S d\tilde{z},
\end{aligned}
$$

which by Ito's lemma leads to dynamics for $c(S,t)$ as

$$
\begin{aligned}
dc &= \left(c_t + (\mu + \sigma\beta)Sc_S + \frac{1}{2}\sigma^2 S^2 c_{SS} \right) dt + \sigma S c_S d\tilde{z} \\
&= \left(c_t + \mu S c_S + \frac{1}{2}\sigma^2 S^2 c_{SS} \right) dt + \sigma S c_S (d\tilde{z} + \beta dt) \\
&= \left(c_t + \mu S c_S + \frac{1}{2}\sigma^2 S^2 c_{SS} \right) dt + \sigma S c_S (d\psi).
\end{aligned}
$$

Note that the factor does not directly enter into the factor model for the risk-free asset B, so it remains unchanged. Placing this information back in a tradable table gives

$$
\begin{array}{c}
\text{Prices} \quad\quad \text{Value Change Factor Models} \\
\hline
\begin{bmatrix} B \\ S \\ c \end{bmatrix} \quad d\begin{bmatrix} B \\ S \\ c \end{bmatrix} = \begin{bmatrix} r_0 B \\ \mu S \\ \left(c_t + \mu S c_S + \frac{1}{2}\sigma^2 S^2 c_{SS}\right) \end{bmatrix} dt + \begin{bmatrix} 0 \\ \sigma S \\ \sigma S c_S \end{bmatrix} d\psi.
\end{array}
$$

Thus, we see that dz and $d\psi = d\tilde{z} + \beta dt$ are consistent in that they produce the same $\mathcal{A}$ and $\mathcal{B}$ factor model coefficients in a tradable table!

The upshot is that we can replace any Brownian factor dz by a Brownian plus a drift $d\tilde{z} + \beta dt$, and the same absence of arbitrage prices will result!

10.1.1.2   Changing the Variance Does Not Work

Let's try new factors given by the replacement

$$dz \to d\psi = \eta d\tilde{z}. \tag{10.5}$$

That is, we will replace the original Brownian factor by a Brownian with a different variance. Again, the new factor should be considered to be the entire term $d\psi = \eta d\tilde{z}$ and not just $d\tilde{z}$. We can ask whether this factor $d\psi = \eta d\tilde{z}$ is consistent with the original dz. We have

$$
\begin{aligned}
dS &= \mu S dt + \sigma S \, (d\psi) \\
&= \mu dt + \sigma S (\eta d\tilde{z}),
\end{aligned}
$$

and then by Ito's lemma for $c(S,t)$,

$$
\begin{aligned}
dc &= \left(c_t + \mu S c_S + \frac{1}{2}\sigma^2\eta^2 S^2 c_{SS} \right) dt + \sigma S c_S \, (\eta d\tilde{z}) \\
&= \left(c_t + \mu S c_S + \frac{1}{2}\sigma^2\eta^2 S^2 c_{SS} \right) dt + \sigma S c_S (d\psi).
\end{aligned}
$$

Gathering this in a tradable tables gives

Prices   Value Change Factor Models

$$
\begin{bmatrix} B \\ S \\ c \end{bmatrix}
\quad
d \begin{bmatrix} B \\ S \\ c \end{bmatrix} =
\begin{bmatrix} r_0 B \\ \mu S \\ \left(c_t + \mu S c_S + \frac{1}{2}\eta^2\sigma^2 S^2 c_{SS} \right) \end{bmatrix} dt +
\begin{bmatrix} 0 \\ \sigma S \\ \sigma S c_S \end{bmatrix} d\psi.
$$

Thus, we see that the drift of c has changed, which means that $\mathcal{A}$ is changed when we change the factor to another Brownian with a different variance. Therefore, this substitution is not acceptable and will not produce absence of arbitrage prices consistent with the original factor dz.

Similar to changing the variance of a single Brownian factor, in the case of multiple Brownian factors changes in the correlations or the covariance structure are not allowed. The above results lead us to the following principle.

($\bigstar$) Arbitrage Invariance Principle for Brownian Motion
If a set of prices $\mathcal{P}$ is absence of arbitrage under Brownian factors $dz \in \mathbb{R}^n$ with $E[dz dz^T] = \Sigma dt$, then $\mathcal{P}$ is also absence of arbitrage if the factors dz are replaced by $dz \to d\psi = d\tilde{z} + \beta dt$ with $\beta \in \mathbb{R}^n$ arbitrary and where $d\tilde{z} \in \mathbb{R}^n$ is a vector of Brownian factors with $E[d\tilde{z} d\tilde{z}^T] = \Sigma dt$.

10.1.2   Poisson Factors

We can also explore whether there exist changes in Poisson factors that preserve the tradable table factor model structure.

10.1.2.1 Changing the Intensity Works

Assume that the original factor is a Poisson process $d\pi(t; \alpha)$ where $\alpha > 0$ is the intensity. Let's consider a simple geometric Poisson motion model of a stock driven by $d\pi(t; \alpha)$, and use Ito's lemma to obtain the factor model for a derivative c on the stock. This leads to

$$
\begin{aligned}
dS &= \mu S^- dt + (\sigma - 1)S^- d\pi(t; \alpha), & (10.6) \\
dc &= (c_t + \mu S^- c_S)dt + (c(\sigma S^-) - c(S^-))d\pi(t; \alpha). & (10.7)
\end{aligned}
$$

Next, let's consider changing the factor to a Poisson with an altered intensity $d\psi = d\tilde{\pi}(t; \alpha + \beta)$ where $\alpha + \beta > 0$. This case results in

$$
\begin{aligned}
dS &= \mu S^- dt + (\sigma - 1)S^- d\tilde{\pi}(t; \alpha + \beta) \\
dc &= (c_t + \mu S^- c_S)dt + (c(\sigma S^-) - c(S^-))d\tilde{\pi}(t; \alpha + \beta)
\end{aligned}
$$

and we see that the $\mathcal{A}$ and $\mathcal{B}$ factor model representations remain the same under $d\psi = d\tilde{\pi}(t; \alpha + \beta)$. This indicates that this substitution of the factors is allowable, and will not disturb absence of arbitrage prices.

10.1.2.2 Adding a Drift Does Not Work

Again assume that the original factor is a Poisson process $d\pi(t; \alpha)$ giving rise to the factor models in (10.6) and (10.7). However, this time let's consider changing the factor by adding a drift, $d\pi(t; \alpha) \to d\psi = d\tilde{\pi}(t; \alpha) + \eta dt$. This will lead to a stock model of

$$
\begin{aligned}
dS &= \mu S^- dt + (\sigma - 1)S^- d\psi \\
&= (\mu S^- + (\sigma - 1)\eta S^-)dt + (\sigma - 1)S^- d\tilde{\pi}(t; \alpha).
\end{aligned}
$$

Applying Ito's lemma to $c(S, t)$ would give

$$
\begin{aligned}
dc &= (c_t + (\mu S^- + (\sigma - 1)\eta S^-)c_S)dt + (c(\sigma S^-) - c(S^-))d\tilde{\pi}(t; \alpha) \\
&= (c_t + (\mu S^- + (\sigma - 1)\eta S^-)c_S - (c(\sigma S^-) - c(S^-))\eta)dt + (c(\sigma S^-) - c(S^-))d\psi,
\end{aligned}
$$

and we see that under the factor $d\psi$, the drift term of c differs from that in equation (10.7). Thus, there is no way to recover the original $\mathcal{A}$ and $\mathcal{B}$ representations under $d\psi$, and this is not an allowable substitution for a Poisson factor.

These results lead us to the following arbitrage invariance principle for Poisson factors.

(★) Arbitrage Invariance Principle for Poisson

If a set of prices $\mathcal{P}$ is absence of arbitrage under a Poisson factor $d\pi(\alpha)$, then $\mathcal{P}$ is also absence of arbitrage under a different Poisson factor $d\psi = d\tilde{\pi}(\alpha + \beta)$ with intensity $\alpha + \beta > 0$.

With these arbitrage invariance principles in hand, we can return to our hypothesis and see whether, by an arbitrage invariant substitution of the factors, we can make pricing of derivatives a little easier.

10.2   RISK NEUTRAL REPRESENTATIONS

The Brownian and Poisson arbitrage invariance principles tell us that it is okay to alter or replace the factors in certain ways. Why is this helpful? Because in some situations it is easier to price a derivative if we replace the original factors by new factors. In fact, in this section we show that we can always replace the factors and put all tradables in what we will call a risk neutral representation. The fact that we can do this will ultimately lead us to the risk neutral pricing principle. The basic idea is that for any pricing problem, if no arbitrage exists, we can replace the factors to create the risk neutral representation which leads to a simplified pricing formula. But we are getting ahead of ourselves. First, let's see what the risk neutral representation is.

10.2.1   Risk Neutrality under Brownian Factors

Consider the tradable table with Brownian factors dz,

$$
\begin{array}{c|c}
\text{Prices} & \text{Value Change Factor Models} \\
-\,-\,-\,- & -\,-\,-\,-\,-\,-\,-\,-\,-\,-\,-\,-\,-\,- \\
[\,\mathcal{P}\,] & d[\,\mathcal{V}\,] = [\,\mathcal{A}\,]\,dt + [\,\mathcal{B}\,]\,dz,
\end{array}
$$

with the accompanying Price APT absence of arbitrage condition, $\mathcal{A} = \mathcal{P}\lambda_0 + \mathcal{B}\lambda$. We can substitute the Price APT back into the tradable table to obtain

$$
\begin{array}{c|c}
\text{Prices} & \text{Value Change Factor Models} \\
-\,-\,-\,- & -\,-\,-\,-\,-\,-\,-\,-\,-\,-\,-\,-\,-\,- \\
[\,\mathcal{P}\,] & d[\,\mathcal{V}\,] = [\,\mathcal{P}\lambda_0 + \mathcal{B}\lambda\,]\,dt + [\,\mathcal{B}\,]\,dz.
\end{array}
\tag{10.8}
$$

Finally, we group terms differently to yield

$$
\begin{array}{c|c}
\text{Prices} & \text{Value Change Factor Models} \\
-\,-\,-\,- & -\,-\,-\,-\,-\,-\,-\,-\,-\,-\,-\,-\,-\,- \\
[\,\mathcal{P}\,] & d[\,\mathcal{V}\,] = [\,\mathcal{P}\lambda_0\,]\,dt + [\,\mathcal{B}\,]\,(dz + \lambda dt).
\end{array}
$$

Now, by the arbitrage invariance principle for Brownian motion, we are allowed to change the drift of the factors and use the replacement $dz \to d\psi = d\tilde{z} - \lambda dt$ which leads to

$$
\begin{array}{c|c}
\text{Prices} & \text{Value Change Factor Models} \\
-\,-\,-\,- & -\,-\,-\,-\,-\,-\,-\,-\,-\,-\,-\,-\,-\,- \\
[\,\mathcal{P}\,] & d[\,\mathcal{V}\,] = [\,\mathcal{P}\lambda_0\,]\,dt + [\,\mathcal{B}\,]\,d\tilde{z}.
\end{array}
$$

This is the risk neutral representation. It states that we can find a replacement set of Brownian factors $d\tilde{z}$ so that all tradables have a drift rate equal to the market price of time λ_0 (which is the risk-free rate or short rate) multiplied by their price. Moreover, absence of arbitrage prices under this representation will be the same as under the original representation using the actual Brownian factors dz.

Note that to obtain this risk neutral representation, the factors were replaced by new factors that contained the market prices of risk! Let's formalize this notion of a risk neutral representation.

(★) **Risk Neutral Representation under Brownian Factors**

Consider the following tradable table,

$$\begin{array}{c|c} Prices & Value\ Change\ Factor\ Models \\ ----\ | & ---------------- \\ [\ \mathcal{P}\ ] & d[\ \mathcal{V}\ ] = [\ \mathcal{A}\ ]\,dt + [\ \mathcal{B}\ ]\,dz, \end{array}$$

and assume that no arbitrage exists. Then, under arbitrage invariant substitutions of the factors $dz \rightarrow d\psi = d\tilde{z} - \lambda dt$, the following tradable table will produce the same absence of arbitrage prices:

$$\begin{array}{c|c} Prices & Value\ Change\ Factor\ Models \\ ----\ | & --------------- \\ [\ \mathcal{P}\ ] & d[\ \mathcal{V}\ ] = [\ \mathcal{P}\lambda_0\ ]\,dt + [\ \mathcal{B}\ ]\,d\tilde{z}. \end{array} \qquad (10.9)$$

This is called the risk neutral representation *because every tradable has a drift rate equal to the market price of time λ_0 multiplied by its price, regardless of how risky it really is.*

10.2.2 Market Price of Risk Interpretation

There is another way to interpret the risk neutral representation. It corresponds to a market in which the market prices of risk for the Brownian factors are equal to zero. To see this, take another look at equation (10.8) and set $\lambda = 0$. This means two things.

First, it means that there is no arbitrage in this market since there is a λ (specifically $\lambda = 0$) that satisfies the Price APT equation for all tradables. Second, it means that the drift term of every tradable will be equal to $\lambda_0\mathcal{P}$, which is the same as the risk neutral representation.

Now, you may protest that in the "real" market, the market prices of risk are not equal to zero! Moreover, you may be absolutely correct! However, the risk neutral representation says that if a world existed where the market prices of risk were, in fact, equal to zero, you would have exactly the same absence of arbitrage prices for derivative securities. Thus, why not pretend that the market prices of risk are zero if that makes pricing easier. This is one way to interpret risk neutral pricing.

In fact, to carry this line of thinking even further, we could pretend that the market prices of risk were other numbers as well, not just zero, and still arrive at the same absence of arbitrage prices for derivatives! It turns out that this is not only correct thinking, but in some cases it is actually quite useful. However, in "most" cases, choosing the market prices of risk equal to zero leads to the "simplest" pricing formula. At this point, I am drifting far afield of the topic at hand. So, to wrap up, let's just note that the risk neutral representation is the same as assuming that the market prices of risk for Brownian factors are equal to zero.

10.2.3 Risk Neutrality under Poisson Factors

Consider the tradable table driven by Poisson factors $d\pi(t,\alpha)$,

$$\text{Prices} \quad \text{Value Change Factor Models}$$
$$[\,\mathcal{P}\,] \quad d[\,\mathcal{V}\,] = [\,\mathcal{A}\,]\,dt + [\,\mathcal{B}\,]\,d\pi(t;\alpha),$$

where the Price APT absence of arbitrage condition is

$$\mathcal{A} = \mathcal{P}\lambda_0 + \mathcal{B}\lambda. \tag{10.10}$$

Before proceeding, let's think about the market price of risk for a Poisson factor $d\pi$. Since a Poisson process either does nothing or jumps up by 1, it is always good to hold a positive amount of a Poisson factor. All the risk is on the upside. On the other hand, being short a Poisson factor is adding real (downside) risk. Thus, we would expect the market price of risk for a Poisson factor to be negative. That is, if you are short a Poisson factor ($\mathcal{B}$ is negative), then you should be rewarded for taking on that risk, and a negative market price of risk would reflect that.

Using this logic, let's assume that $\lambda < 0$ for a Poisson factor. I don't like dealing with a negative quantity, so let's define $\lambda' = -\lambda$ and rewrite the absence of arbitrage condition in terms of λ' as

$$\begin{aligned}\mathcal{A} &= \mathcal{P}\lambda_0 + \mathcal{B}\lambda \\ &= \mathcal{P}\lambda_0 + \mathcal{B}(-\lambda') \\ &= \mathcal{P}\lambda_0 - \mathcal{B}\lambda'\end{aligned}$$

where $\lambda' > 0$. We can substitute this into the tradable table to obtain

$$\text{Prices} \quad \text{Value Change Factor Models}$$
$$[\,\mathcal{P}\,] \quad d[\,\mathcal{V}\,] = [\,\mathcal{P}\lambda_0 - \mathcal{B}\lambda'\,]\,dt + [\,\mathcal{B}\,]\,d\pi(t;\alpha).$$

Now, by the arbitrage invariance principle for Poisson factors, changes in the intensity are allowed and we can use the replacement $d\pi(t;\alpha) \to d\psi = d\tilde{\pi}(t;\alpha+\beta)$.

Let's choose $\beta = \lambda' - \alpha$ so that $d\psi = d\tilde{\pi}(t; \lambda')$ (here is where it is important that $\lambda' > 0$, so that it can be an intensity!), which leads to

$$
\begin{array}{c|c}
\text{Prices} & \text{Value Change Factor Models} \\
\text{----} & \text{---------------} \\
[\; \mathcal{P}\;] & d\,[\; \mathcal{V}\;] = [\; \mathcal{P}\lambda_0 - \mathcal{B}\lambda'\;]\,dt + [\;\mathcal{B}\;]\,d\tilde{\pi}(t; \lambda').
\end{array}
$$

The final step is to compensate the Poisson process. That is, we shift part of the drift and associate it with the Poisson process so that the random factor term has mean zero:

$$
\begin{array}{c|c}
\text{Prices} & \text{Value Change Factor Models} \\
\text{----} & \text{--------------} \\
[\; \mathcal{P}\;] & d\,[\; \mathcal{V}\;] = [\; \mathcal{P}\lambda_0\;]\,dt + [\;\mathcal{B}\;]\,(d\pi(t; \lambda') - \lambda' dt).
\end{array}
$$

Since the random factor term now has zero mean, we see that the drift rate for each tradable is equal to the market price of time λ_0 multiplied by its price. This mirrors what we saw in the Brownian case, and is the risk neutral representation under Poisson factors. Note that in this case, the new intensity of the factors in the risk neutral representation is equal to (minus) the market price of risk. Let's formalize this.

(★) Risk Neutral Representation for Poissons
Consider the following tradable table,

$$
\begin{array}{c|c}
\textit{Prices} & \textit{Value Change Factor Models} \\
\text{----} & \text{--------------} \\
[\; \mathcal{P}\;] & d\,[\; \mathcal{V}\;] = [\; \mathcal{A}\;]\,dt + [\;\mathcal{B}\;]\,d\pi(t; \alpha)
\end{array}
$$

and assume that no arbitrage exists. Then, under an arbitrage invariant substitutions of the factor $d\pi(t; \alpha) \to d\psi = d\tilde{\pi}(t; \lambda')$, where $\lambda' = -\lambda$ is minus the market price of risk, the following tradable table will produce the same absence of arbitrage prices:

$$
\begin{array}{c|c}
\textit{Prices} & \textit{Value Change Factor Models} \\
\text{----} & \text{-------------} \\
[\; \mathcal{P}\;] & d\,[\; \mathcal{V}\;] = [\; \mathcal{P}\lambda_0\;]\,dt + [\;\mathcal{B}\;]\,(d\tilde{\pi}(t; \lambda') - \lambda' dt). \quad (10.11)
\end{array}
$$

Equation (10.11) is called the risk neutral representation *because every tradable has a drift rate equal to the market price of time multiplied by its price, regardless of how risky it really is.*

10.2.4 Market Price of Risk under Poisson Factors

Here, we see a difference between the Brownian factor case and the Poisson factor case. In the Brownian case, the risk neutral representation could be achieved simply by setting the market price of risk equal to zero for the Brownian factor. In the Poisson case, the risk neutral representation involves setting the intensity of the Poisson factor equal to minus the actual market price of risk $\lambda' = -\lambda$. Thus, in the Poisson case, risk neutral pricing doesn't exactly allow us to bypass the determination of the actual market price of risk. However, the resulting tradable table can be interpreted again as a world where there is no reward (i.e., the market price of risk is zero) for the new compensated Poisson factor $(d\tilde{\pi}(t; \lambda') - \lambda' dt)$.

10.3 PRICING AS AN EXPECTATION

The risk neutral representation leads to a powerful new approach to absence of arbitrage pricing because instead of using the real tradable table, we can use the risk neutral representation. The key idea is to take advantage of the fact that the expected rate of return of all tradables is equal to the market price of time. Let's see how it works for the Brownian case.

Consider any tradable under the risk neutral representation, where by equation (10.9) we have

$$dV = \lambda_0 \mathcal{P} dt + \mathcal{B} d\tilde{z}. \tag{10.12}$$

Let's assume that we are not dealing with a futures contract so that $V = \mathcal{P}$. Thus,

$$d\mathcal{P} = \lambda_0 \mathcal{P} dt + \mathcal{B} d\tilde{z}, \tag{10.13}$$

which indicates that the expected rate of return of $\mathcal{P}$ is λ_0 when we use the risk neutral factors $d\tilde{z}$. Now, via Ito's lemma one can verify that

$$d\left(e^{-\int_0^t \lambda_0(s)ds}\mathcal{P}\right) = e^{-\int_0^t \lambda_0(s)ds}\mathcal{B}d\tilde{z}. \tag{10.14}$$

Taking expectations of both sides gives

$$d\mathbb{E}^*\left(e^{-\int_0^t \lambda_0(s)ds}\mathcal{P}\right) = 0 \tag{10.15}$$

since $\tilde{z}$ is a Brownian motion and the increments of the Brownian motion have zero mean, $\mathbb{E}[d\tilde{z}] = 0$. (We also switched the $d(\cdot)$ and the expectation on the left side of the equation.) Finally, integration of both sides shows that

$$\mathcal{P}(0) = \mathbb{E}^*\left[e^{-\int_0^t \lambda_0(s)ds}\mathcal{P}(t)\right]. \tag{10.16}$$

We have arrived at the risk neutral pricing formula.

(★) The Risk Neutral Pricing Formula
Absence of arbitrage prices are given by

$$\mathcal{P}(0) = \mathbb{E}^*\left[e^{-\int_0^t \lambda_0(s)ds}\mathcal{P}(t)\right] \tag{10.17}$$

where the expectation $\mathbb{E}^*[\cdot]$ is taken under the risk neutral representation and λ_0 is the market price of time.

Note what the risk neutral pricing formula (10.17) says. We know that under the risk neutral representation, all tradables have an expected rate of return equal to λ_0. So, the formula indicates you should take the price at some time in the future $\mathcal{P}(t)$ (it makes sense to choose t as an expiration time if you want to price an option) and remove the effect of the λ_0 rate of return that has accumulated up to time t by discounting by $e^{-\int_0^t \lambda_0(s)ds}$. Finally, since you are in a risk neutral world and risk doesn't matter to you, price by taking the expected value $\mathbb{E}^*[\cdot]$. That is it!

In a similar manner, which I won't detail here, this risk neutral pricing principle applies in the Poisson case as well. Thus, this is a new pricing point of view that follows from the factor approach!

The route to the risk neutral pricing formula that I have taken here is not standard. The foundations of risk neutral pricing and related concepts were laid out in [23, 24] in terms of probability/measure theory and martingales. Those seeking a more traditional and rigorous treatment are referred to the book of Duffie [16]. Nevertheless, the benefit of the approach given in this chapter is that risk neutral pricing is revealed as a natural consequence of the factor model approach.

The presentation has been a little abstract to this point, so let's apply risk neutral pricing to some concrete examples before wrapping up this chapter.

10.4 APPLICATIONS OF RISK NEUTRAL PRICING

In this section we will see how risk neutral pricing is used in a couple of familiar situations. But first, let's outline how it is applied in general.

10.4.1 How to Apply Risk Neutral Pricing

To use risk neutral pricing, one approach is to start with the real model (i.e., the model you believe is "real") of the underlying variables and tradables, and then apply an arbitrage invariant factor substitution to bring it to the risk neutral representation. However, it is quite common to just directly write a risk neutral representation for all variables. This includes the factors, underlying variables, and tradables. In the risk neutral tradable table, the drift of every asset is equal to the market price of time (i.e., either the risk-free rate or the short rate of interest corresponding to a money market account) multiplied by its price. As long as all the marketed tradables are correctly priced using the risk neutral pricing formula (10.17), then there is no arbitrage under this risk neutral representation.

Additionally, and this is the important part, there is also no arbitrage under any arbitrage invariant substitution of the factors. As long as you believe that the real model can be created from the risk neutral representation via an arbitrage invariant factor substitution, then the prices that you compute are also absence of arbitrage for your real market, and you are good to go.

Now, when considering whether your real market is within an arbitrage invariant factor substitution of your risk neutral representation, you must recall how an arbitrage invariant substitution is allowed to change the factors. For example, when dealing with Brownian factors, arbitrage invariant factor substitutions aren't allowed to change the volatility or correlation structure of the factors. Thus, roughly speaking, at a minimum you must use the real volatilities and correlations in your risk neutral representation, otherwise it will definitely not "cover" the real market.

To summarize, here are the basic steps in applying the risk neutral pricing approach,

1. Create a risk neutral representation of all your variables, including the factors, underlying variables, and tradables. Recall that for it to be a risk neutral representation, the drift rate of each tradables must equal the market price of time λ_0 multiplied by its price. Moreover, this risk neutral representation must be able to cover what you believe is the real model via an arbitrage invariant substitution of factors.

2. Make sure that your risk neutral representation correctly prices all marketed tradables using equation (10.17). This is the calibration phase. If one first specifies just the *form* of the risk neutral representation, without assigning specific parameter values, then this is the step in which specific parameter values would be selected in order to match the actual prices of the marketed tradables.

3. Use the risk neutral pricing formula (10.17) to price derivative securities.

In the examples that follow, for clarity I will emphasize the relationship between the real model and the risk neutral model. But, as mentioned previously, it is not uncommon to see this aspect of risk neutral pricing glossed over.

10.4.2 Black–Scholes

Let's see how risk neutral pricing applies in the Black–Scholes setup. Recall that we believe the real model for the bond, stock, and derivative is

$$
\begin{array}{cc}
\text{Prices} & \text{Value Change Factor Models} \\
\end{array}
$$

$$
\begin{bmatrix} B \\ S \\ \hline c \end{bmatrix} \quad d \begin{bmatrix} B \\ S \\ \hline c \end{bmatrix} = \begin{bmatrix} r_0 B \\ \mu S \\ \hline c_t + \mu S c_S + \frac{1}{2}\sigma^2 S^2 c_{SS} \end{bmatrix} dt + \begin{bmatrix} 0 \\ \sigma S \\ \hline \sigma S c_S \end{bmatrix} dz.
$$

However, according to the risk neutral pricing principle, we have the same absence of arbitrage prices if we set the drifts of all these tradables to the risk-free rate (market price of time) multiplied by their prices, since this can be achieved by an arbitrage invariant substitution of the factors. In particular, using $d\psi = d\tilde{z} - \lambda dt$ where $\lambda = \frac{\mu - r_0}{\sigma}$, and substituting this for the factor dz will achieve this transformation to the risk neutral representation, resulting in:

$$
\begin{array}{cc}
\text{Prices} & \text{Value Change Factor Models} \\
\end{array}
$$

$$
\begin{bmatrix} B \\ S \\ \hline c \end{bmatrix} \quad d \begin{bmatrix} B \\ S \\ \hline c \end{bmatrix} = \begin{bmatrix} r_0 B \\ r_0 S \\ \hline r_0 c \end{bmatrix} dt + \begin{bmatrix} 0 \\ \sigma S \\ \hline \sigma S c_S \end{bmatrix} d\tilde{z}.
$$

Now, we can apply the risk neutral pricing formula (10.17) to the call option $c(S,t)$,

$$
c(S(0),0) = \mathbb{E}^* \left[e^{-r_0 T} c(S(T),T) \right], \tag{10.18}
$$

where the expectation $\mathbb{E}^*[\cdot]$ is taken under the risk neutral representation. But, if T is expiration, then we know that $c(S(T),T) = \max\{S(T) - K, 0\}$. Substituting this into (10.18) leads to

$$
c(S(0),0) = e^{-r_0 T} \mathbb{E}^* \left[\max\{S(T) - K, 0\} \right], \tag{10.19}
$$

where $e^{-r_0 T}$ was pulled out of the expectation because it is not random. This expectation can actually be evaluated explicitly, leading to the Black–Scholes formula. This calculation is shown in the following example.

Example 10.1 (Risk Neutral Derivation of the Black–Scholes Formula)
We can use the risk neutral pricing formula to derive the Black–Scholes formula. The expectation in equation (10.19) can be written as

$$
\begin{aligned}
\mathbb{E}^*[\max\{S(T) - K, 0\}] &= \mathbb{E}^* \left[(S(T) - K)\mathbf{1}_{\{S(T) \geq K\}} \right] \\
&= \mathbb{E}^* \left[S(T)\mathbf{1}_{\{S(T) \geq K\}} \right] - K\mathbb{E}^* \left[\mathbf{1}_{\{S(T) \geq K\}} \right]
\end{aligned}
$$

where $\mathbf{1}_{\{S(T) \geq K\}}$ is the indicator function of the set $\{S(T) \geq K\}$. That is, it is the function that takes the value 1 when $S(T) \geq K$ and 0 otherwise.

Let's take the two terms in the above expectation, $\mathbb{E}^*[S(T)\mathbf{1}_{\{S(T) \geq K\}}]$ and $K\mathbb{E}^*[\mathbf{1}_{\{S(T) \geq K\}}]$, one at a time. The second term is easier to handle first. Note that

$$\mathbb{E}^* \left[\mathbf{1}_{\{S(T) \geq K\}}\right] = \mathbb{P}^*(S(T) \geq K)$$

where $\mathbb{P}^*(\cdot)$ denotes the risk neutral probability. To calculate this probability, we can explicitly write $S(T)$ under the risk neutral representation in terms of the driving Brownian motion as

$$\mathbb{P}^*(S(T) \geq K) = \mathbb{P}^* \left(S(0)e^{(r_0 - 0.5\sigma^2)T + \sigma \tilde{z}(T)} \geq K\right) = \mathbb{P}^* \left(\frac{\tilde{z}(T)}{\sqrt{T}} \geq -d_2\right)$$

where

$$d_2 = \frac{\ln\left(\frac{S(0)}{K}\right) + (r_0 - \frac{1}{2}\sigma^2)T}{\sigma\sqrt{T}}.$$

Now, since $\tilde{z}$ is a Brownian motion, at time T it is normally distributed with mean 0 and standard deviation $\sqrt{T}$. Thus, $\frac{\tilde{z}(T)}{\sqrt{T}} \sim \mathcal{N}(0,1)$ is a standard normal random variable (mean 0 and variance 1). By the symmetry of the standard normal, we can compute the above probability as

$$\mathbb{P}^* \left(\frac{\tilde{z}(T)}{\sqrt{T}} \geq -d_2\right) = \Phi(d_2)$$

where $\Phi(\cdot)$ is the cumulative distribution of the standard normal. This takes care of the second term.

The first term, $\mathbb{E}^*[S(T)\mathbf{1}_{\{S(T) \geq K\}}]$, can be computed by writing out the expectation explicitly in terms of the density function of $\tilde{z}(T)$ as

$$
\begin{aligned}
\mathbb{E}^*[(S(T)\mathbf{1}_{\{S(T) \geq K\}}] &= \mathbb{E}^* \left[S(0)e^{(r_0 - 0.5\sigma^2)T + \sigma\tilde{z}(T)}\mathbf{1}_{\left\{\frac{\tilde{z}(T)}{\sqrt{T}} \geq -d_2\right\}}\right] \\
&= \int_{z \geq -\sqrt{T}d_2} S(0)e^{(r_0 - (0.5)\sigma^2)T + \sigma z}\frac{1}{\sqrt{2\pi T}}e^{-\frac{z^2}{2T}}dz \\
&= \int_{z \geq -\sqrt{T}d_2} S(0)e^{r_0 T}\frac{1}{\sqrt{2\pi T}}e^{-\frac{(z - \sigma T)^2}{2T}}dz \\
&= S(0)e^{r_0 T}\int_{z \geq -\sqrt{T}d_2} \frac{1}{\sqrt{2\pi T}}e^{-\frac{(z - \sigma T)^2}{2T}}dz.
\end{aligned}
$$

To continue, we note that the quantity inside the integral is identical to the density function of a normal random variable with mean σT and standard deviation $\sqrt{T}$. Let

$\hat{z}$ be such a random variable. Then the integral can be expressed as the probability,

$$\int_{z \geq -\sqrt{T}d_2} \frac{1}{\sqrt{2\pi T}} e^{-\frac{(z-\sigma T)^2}{2T}} dz = \mathbb{P}\left(\hat{z} \geq -\sqrt{T}d_2\right)$$

$$= \mathbb{P}\left(\frac{\hat{z} - \sigma T}{\sqrt{T}} \geq -d_2 - \sigma\sqrt{T}\right)$$

$$= \Phi(d_2 + \sigma\sqrt{T})$$

where we have used the fact that $\frac{\hat{z}-\sigma T}{\sqrt{T}} \sim \mathcal{N}(0,1)$ is a standard normal random variable. Thus, we can compute everything required for the first term in closed form as well.

By defining

$$d_1 = d_2 + \sigma\sqrt{T} = \frac{\ln\left(\frac{S(0)}{K}\right) + (r_0 + \frac{1}{2}\sigma^2)T}{\sigma\sqrt{T}},$$

we can put all these calculations together resulting in the Black–Scholes closed form solution,

$$c(S(0), 0) = S(0)\Phi(d_1) - Ke^{-r_0 T}\Phi(d_2). \tag{10.20}$$

Thus, we were able to obtain the pricing formula without resorting to partial differential equations!

10.4.3 Poisson Model

Risk neutral pricing applies equally well when we are dealing with Poisson factors. In this section, let's apply it to the Poisson model of Chapter 6, Section 6.4. To be consistent with the approach in Section 10.2.3, I won't compensate the original Poisson factor $d\pi(t; \nu)$ as was done in Section 6.4. The tradable table is

$$
\begin{array}{cc}
\text{Prices} & \text{Value Change Factor Models} \\
- - - - \;\; | & - - - - - - - - - - - - - - \\
\begin{bmatrix} B \\ S^- \\ -- \\ c \end{bmatrix} & d \begin{bmatrix} B \\ S^- \\ -- \\ c \end{bmatrix} = \begin{bmatrix} r_0 B \\ \mu S^- \\ -- \\ c_t + \mu S^- c_S \end{bmatrix} dt + \begin{bmatrix} 0 \\ (k-1)S^- \\ -- \\ c(kS^-) - c(S^-) \end{bmatrix} d\pi(\nu).
\end{array}
$$

and the market price of time is $\lambda_0 = r_0$, while the market price of risk is

$$\lambda = \frac{\mu - r_0}{k - 1}. \tag{10.21}$$

For the risk neutral representation we use minus the market price of risk

$$-\lambda = \lambda' = \frac{r_0 - \mu}{k - 1} \tag{10.22}$$

and alter the intensity of the Poisson process to λ', leading to

Prices   Value Change Factor Models

$$
\begin{bmatrix} B \\ S^- \\ -- \\ c \end{bmatrix} \quad d\begin{bmatrix} B \\ S^- \\ -- \\ c \end{bmatrix} = \begin{bmatrix} r_0 B \\ r_0 S^- \\ -- \\ r_0 c \end{bmatrix} dt + \begin{bmatrix} 0 \\ (k-1)S^- \\ --- \\ c(kS^-) - c(S^-) \end{bmatrix} (d\pi(\lambda') - \lambda' dt).
$$

Now that we are in the risk neutral representation, we can apply the risk neutral pricing formula,

$$
c(S(0),0) = \mathbb{E}^*\big[e^{-r_0 T} c(S(T),T)\big], \tag{10.23}
$$

where T is expiration. If c is a European call option that expires at time T, we can replace $c(S(T),T)$ by its payoff value $c(S(T),T) = \max\{S(T) - K, 0\}$, so the risk neutral pricing formula becomes

$$
c(S(0),0) = e^{-r_0 T} \mathbb{E}^*[\max\{S(T) - K, 0\}] \tag{10.24}
$$

where the term $e^{-r_0 T}$ was pulled out of the expectation because it is not random. Similar to the Black–Scholes case, the expectation in (10.24) can be computed in closed form as

$$
c(S,t) = S\Psi(x,y) - Ke^{-r_0(T)}\Psi(x,y/k) \tag{10.25}
$$

where

$$
\Psi(\alpha, \beta) = \sum_{i=\alpha}^{\infty} \frac{e^{-\beta}\beta^i}{i!}, \quad y = \frac{(r_0 - \mu)kT}{k-1}, \tag{10.26}
$$

and x is the smallest non-negative integer greater than $\frac{\ln(K/S) - \mu(T)}{\ln(k)}$. Details of this calculation are left to you.

10.4.4 Vasicek Single Factor Short Rate Model

To further illustrate the risk neutral approach, let's revisit the Vasicek single factor short rate term structure model of Chapter 7, Section 7.3.2.1. In this model, we had the following classification of variables,

Tradables	Underlying Variables	Factors
$B_0, B(T)$	r_0	dt, dz

with r_0 the short rate process, B_0 the money market account, and $B(T) = B(r_0, t|T)$ representing zero-coupon bonds paying \$1 at maturity T. The short rate was modeled as an Ornstein–Uhlenbeck mean-reverting process,

$$
dr_0 = \kappa(\theta - r_0)dt + b\,dz.
$$

Our tradable table was comprised of the money market account B_0 that earned the instantaneous short rate, and zero-coupon bonds $B(T)$ of maturity T:

$$
\begin{array}{c|c}
\text{Prices} & \text{Value Change Factor Models} \\
---- & ------------- \\
\begin{bmatrix} B_0 \\ B(T) \end{bmatrix} & d\begin{bmatrix} B_0 \\ B(T) \end{bmatrix} = \begin{bmatrix} r_0 B_0 \\ (B_t(T) + aB_r(T) + \frac{1}{2}b^2 B_{rr}(T)) \end{bmatrix} dt + \begin{bmatrix} 0 \\ bB_r(T) \end{bmatrix} dz.
\end{array}
$$

To create the risk neutral representation, we need to find a change of the Brownian factor $dz \to d\psi = d\tilde{z} - \lambda dt$ so that all tradables have a drift rate equal to the short rate multiplied by their price. That is, upon substitution of $d\psi$ for dz in the tradable table above, it becomes

$$
\begin{array}{c|c}
\text{Prices} & \text{Value Change Factor Models} \\
---- & ------------- \\
\begin{bmatrix} B_0 \\ B(T) \end{bmatrix} & d\begin{bmatrix} B_0 \\ B(T) \end{bmatrix} = \begin{bmatrix} r_0 B_0 \\ r_0 B(T) \end{bmatrix} dt + \begin{bmatrix} 0 \\ bB_r(T) \end{bmatrix} d\tilde{z}.
\end{array}
$$

However, in this case, we can't forget to also swap out the factor in the underlying variable! That is, under the risk neutral representation, the dynamics of the underlying variable short rate also changes to

$$dr_0 = \kappa(\theta - r_0)dt + b(d\tilde{z} - \lambda dt) = (\kappa(\theta - r_0) - \lambda b)dt + bd\tilde{z}, \tag{10.27}$$

where, if we recall, λ is the market price of risk for dz. Thus, it seems that we have not escaped the task of determining the market price of risk λ.

However, here is where calibration of the market price of risk is sometimes "hidden" in the risk neutral approach. Instead of parameterizing the short rate dynamics in terms of λ as in (10.27), we write it as

$$dr_0 = \tilde{\kappa}(\tilde{\theta} - r_0)dt + bd\tilde{z} \tag{10.28}$$

where it is assumed that $\tilde{\kappa}(\tilde{\theta} - r_0) = (\kappa(\theta - r_0) - \lambda b)$.

Thus, we can write the risk neutral dynamics in terms of equation (10.28), where the selection of $\tilde{\kappa}$ and $\tilde{\theta}$ replaces the calibration of the market price of risk λ. In particular, the parameters $\tilde{\kappa}$ and $\tilde{\theta}$ must be chosen so that the risk neutral formula applied to the marketed tradable zero-coupon bonds $B(T)$ matches their actual market prices.

More specifically, using the fact that the payoff of a zero-coupon bond of maturity T is $B(r_0(T), T|T) = 1$, the risk neutral pricing formula reduces to

$$B(r_0(0), 0|T) = \mathbb{E}^*\left[e^{-\int_0^T r_0(s)ds} B(r_0(T), T|T)\right] = \mathbb{E}^*\left[e^{-\int_0^T r_0(s)ds}\right]. \tag{10.29}$$

Thus, $\tilde{\kappa}$ and $\tilde{\theta}$ are properly calibrated when this formula agrees with the market prices for the bonds $B(T)$. A numerical example illustrating this calibration is given next.

Example 10.2 (Calibrating the Risk Neutral Vasicek Model)
To calibrate the risk neutral Vasicek model to zero-coupon bonds, we need to be able to evaluate the risk neutral formula

$$B(r_0(0), 0|T) = \mathbb{E}^*\left[e^{-\int_0^T r_0(s)ds}\right],$$

where r_0 is governed by equation (10.28). It turns out that this expectation can be computed in closed form as

$$
\begin{aligned}
B(r_0(0), 0|T) &= \exp(x(0)r_0(0) + y(0)), & (10.30) \\
x(t) &= \frac{\exp(\tilde{\kappa}(t-T)) - 1}{\tilde{\kappa}}, & (10.31) \\
y(t) &= \left(-\tilde{\theta} + \frac{b^2}{2\tilde{\kappa}^2}\right)(x(t) - (t-T)) - \frac{b^2 x(t)^2}{4\tilde{\kappa}}. & (10.32)
\end{aligned}
$$

(Note that this is essentially the same formula that appears in equations (7.26) through (7.28) of Chapter 7, Section 7.3.2.1 on the Vasicek Model.)

Let the current short rate be $r_0(0) = 0.05$ and take $b = 0.02$ in the risk neutral short rate model of equation (10.28).

Consider two zero-coupon bonds of maturity $T = 1$ year and $T = 2$ years with market prices of $B(0|1) = 0.9491$ and $B(0|2) = 0.8975$, respectively. To calibrate the Vasicek risk neutral short rate model of (10.28), the parameters $\tilde{\theta}$ and $\tilde{\kappa}$ must be chosen so that the formula in (10.30) matches the market prices of these two bonds. In this case, one may verify that using the values $\tilde{\theta} = 0.07$ and $\tilde{\kappa} = 0.25$ results in the correct prices. Thus, the calibrated risk neutral short rate model of (10.28) is given by

$$dr_0 = 0.25(0.07 - r_0)dt + 0.02d\tilde{z}.$$

Finally, to price a derivative security that depends on the short rate $c(r_0(t), t)$, one would simply need to evaluate (whether in closed form or via numerical methods) the risk neutral pricing formula

$$c(r_0(0), 0) = \mathbb{E}^*\left[e^{-\int_0^T r_0(s)ds} c(r_0(T), T)\right] \qquad (10.33)$$

under the calibrated short rate model for r_0, with $c(r_0(T), T)$ being the payoff value of the derivative at its expiration T.

10.4.5 Heath–Jarrow–Morton Calibration

As a final example, we show how easy risk neutral pricing makes the calibration of the Heath–Jarrow–Morton model of Chapter 7, Section 7.6. In that model, the underlying variables are given by the instantaneous forward rates that follow

$$df(t|s) = a(s)dt + b(s)dz, \qquad (10.34)$$

where the instantaneous short rate of interest is given by $r_0(t) = f(t|t)$. Additionally, by equation (7.122) in Appendix 7.9, the tradables are zero-coupon bonds that follow

$$dB(t|T) = B(t|T)\left[\tilde{\mu}(T)dt - \left(\int_t^T b(s)ds\right)dz\right]$$

with

$$\tilde{\mu}(T) = \left(r_0(t) - \int_t^T a(s)ds + \frac{1}{2}\int_t^T\int_t^T b(s)b(r)drds\right). \tag{10.35}$$

The tradable table was then given by the zero-coupon bonds and a money market account:

$$
\begin{array}{cc}
\text{Prices} & \text{Value Change Factor Models} \\
---- \ | & ----------- \\
\begin{bmatrix} B_0 \\ B(t|T) \end{bmatrix} & d\begin{bmatrix} B_0 \\ B(t|T) \end{bmatrix} = \begin{bmatrix} r_0(t)B_0 \\ \tilde{\mu}(T)B(t|T) \end{bmatrix}dt + \begin{bmatrix} 0 \\ -B(t|T)\left(\int_t^T b(s)ds\right) \end{bmatrix}dz.
\end{array}
$$

You may recall from Appendix 7.9 that this was a bit messy to deal with (which is why it was relegated to an appendix!). So, let's start over but with the risk neutral perspective in mind.

To move to the risk neutral representation, we look for a new factor $d\psi = d\tilde{z} - \lambda dt$ to substitute for dz which will lead to all tradables earning the instantaneous short rate of interest, $r_0(t) = f(t|t)$. That is, the risk neutral tradable table will be

$$
\begin{array}{cc}
\text{Prices} & \text{Value Change Factor Models} \\
---- \ | & ----------- \\
\begin{bmatrix} B_0 \\ B(t|T) \end{bmatrix} & d\begin{bmatrix} B_0 \\ B(t|T) \end{bmatrix} = \begin{bmatrix} r_0(t)B_0 \\ r_0(t)B(t|T) \end{bmatrix}dt + \begin{bmatrix} 0 \\ -B(t|T)\left(\int_t^T b(s)ds\right) \end{bmatrix}d\tilde{z}.
\end{array}
$$

But, we cannot forget the effect of the change of the factor from dz to $d\psi = d\tilde{z} - \lambda dt$ on the underlying variables, which are the instantaneous forward rates.

However, in this case, there is a nice shortcut to determine how the move to the risk neutral representation affects the dynamics of the forward rates. The key is that the relationship between the tradable zero-coupon bonds $B(t|T)$ and the instantaneous forward rates is explicitly given by

$$B(t|T) = e^{-\int_t^T f(t|s)ds},$$

or, upon solving for $f(t|T)$,

$$f(t|T) = -\frac{\partial}{\partial T}\ln B(t|T). \tag{10.36}$$

Thus, we can use Ito's lemma, coupled with the dynamics of $B(t|T)$ in the risk neutral representation, to determine what the dynamics of $f(t|T)$ must be in the risk neutral representation.

For simplicity of notation, under the risk netural representation let

$$dB(t|T) = r_0(t)B(t|T)dt + \nu(t|T)B(t|T)d\tilde{z} \tag{10.37}$$

where $\nu(t|T) = -\int_t^T b(s)ds$ or alternatively $b(T) = -\frac{\partial}{\partial T}\nu(t|T)$. Then, applying Ito's lemma to equation (10.36) and allowing for an exchange of $d(\cdot)$ and $\frac{\partial}{\partial T}$ gives

$$df(t|T) = -\frac{\partial}{\partial T}\left(d\left(\ln B(t|T)\right)\right) \tag{10.38}$$

$$= \frac{\partial}{\partial T}\left(\frac{1}{2}\nu^2(t|T) - r_0(t)\right)dt - \frac{\partial}{\partial T}\nu(t|T)d\tilde{z} \tag{10.39}$$

$$= \frac{\partial}{\partial T}\left(\frac{1}{2}\nu^2(t|T)\right)dt - \frac{\partial}{\partial T}\nu(t|T)d\tilde{z}. \tag{10.40}$$

This result can be more conveniently written in terms of $b(T)$ as

$$df(t|T) = \tilde{a}(T)dt + b(T)d\tilde{z}, \tag{10.41}$$

where the drift term $\tilde{a}(T)$ under the risk neutral representation is given by

$$\tilde{a}(T) = \frac{\partial}{\partial T}\left(\frac{1}{2}\nu^2(t|T)\right) = \nu(t|T)\frac{\partial}{\partial T}\nu(t|T) = b(T)\int_t^T b(s)ds. \tag{10.42}$$

The upshot of this is that under the risk neutral representation we can calibrate the HJM model simply by modeling the real volatilities $b(s)$ of the forward rates. All other parameters in the risk neutral representation will follow. The drift of the forward rates $\tilde{a}(s)$ is given by equation (10.42) and the expected rate of return of the tradables is given by the short rate $r_0(t) = f(t|t)$. This shows that calibration becomes quite trivial under the risk neutral representation!

(As a side note, this risk neutral calibration equation (10.42) could have also been obtained simply by setting the market price of risk λ equal to zero in the calibration equation (7.106) of Chapter 7, Section 7.6 where the HJM model was considered.)

To price derivatives, one then applies the risk neutral pricing formula, which in this case takes the form

$$c(0) = \mathbb{E}^*\left[e^{-\int_0^T r_0(s)ds}c(T)\right] = \mathbb{E}^*\left[e^{-\int_0^T f(s|s)ds}c(T)\right],$$

where $c(T)$ is the payoff of a derivative which, presumably, can be expressed as a function of the underlying variables which are the instantaneous forward rates $f(T|s)$. Monte Carlo simulations or other numerical procedures can then be used to evaluate the risk neutral expectation.

10.5 SUMMARY

The point of this chapter was to show that risk neutral pricing is a logical consequence of the factor approach to derivative pricing. The idea was that details of the factors do not seem to appear in the Price APT equations that we used throughout the book. That gave us the idea that perhaps we could change the factors (as long as we didn't disturb the basic factor coefficient structure) and still arrive at the same absence of arbitrage prices. By using a different set of factors (that still preserved the absence of arbitrage prices), in many cases we can simplify calculations. This is the basic notion of risk neutral pricing. In fact, there is quite a bit more that one can do when the full power and generality of the risk neutral approach is explored. But that, my friends, will have to be the subject of another book...

EXERCISES

10.1 Risk Neutral Pricing with Dividends.

Consider a stock that pays a continuous dividend at a rate of q. Show that the expected drift rate of the price of the stock $S(t)$ under the risk neutral representation is $r_0 - q$ where r_0 is the risk-free rate.

10.2 Risk Neutral Pricing with a Futures Contract.

Show that the expected drift rate of a futures price is equal to zero under the risk neutral representation.

10.3 Compute the expected drift rate of a forward price under the risk neutral representation. Make the following assumptions. Under the risk neutral representation, the forward price and the price of a zero-coupon bond with maturity matching expiration of the forward contract follow

$$
\begin{aligned}
dF &= \mu F dt + \sigma F dz_1, \\
dB &= r_0 B dt + \sigma_B B dz_2,
\end{aligned}
$$

where $\mathbb{E}[dz_1 dz_2] = \rho dt$. Recall that the forward price $F(t|T)$ is related to the spot price $S(t)$ and zero-coupon bond price $B(t|T)$ through the equation $S(t) = B(t|T)F(t|T)$. Use the fact that $S(t)$ is tradable and will have rate of return equal to the short rate r_0 under the risk neutral representation to derive what μ must be.

10.4 Verify equation (10.14).

10.5 Consider the Vasicek model. Determine the explicit relationship between the parameters of the short rate process in the risk neutral representation, $\tilde{\kappa}$ and $\tilde{\theta}$ in terms of the real parameters of the short rate process κ, θ, b, when the market price of risk λ is constant. (Hint: Use the equation in the text following equation (10.28) to determine the relationships.)

10.6 Use the fact that the drift rate of the futures price under the risk neutral representation is equal to 0 (see Exercise 10.2) to argue that $f(t|T) = S(t)e^{r_0(T-t)}$ where $f(t|T)$ is the futures price with expiration T and $S(t)$ is the spot price. Assume that the risk-free rate r_0 is constant. (Hint: Use the risk neutral pricing formula for $f(t|T)$ and $S(t)$ along with the fact that $f(T|T) = S(T)$.)

10.7 Use risk neutral arguments to derive the Black–Scholes partial differential equation (6.6) for a derivative of the form $c(S, t)$ on a non-dividend paying stock that follows geometric Brownian motion. Proceed as follows. Write the risk neutral representation for the stock. Next, use Ito's lemma applied to $c(S, t)$ under the risk neutral representation and calculate the drift term. Since this is the under the risk neutral representation, set that drift term equal to $r_0 c$ where r_0 is the risk-free rate. This should be the Black–Scholes equation.

10.8 Repeat Exercise 10.7 for a derivative $c(S, t)$ on a stock S that pays a continuous dividend at a rate of q.

Bibliography

[1] L. Bergomi. *Stochastic Volatility Modeling.* Chapman & Hall/CRC, 2016.

[2] F. Black. The Pricing of Commodity Contracts. *Journal of Financial Economics,* 3:167–179, 1976.

[3] F. Black, E. Derman, and W. Toy. A One-Factor Model of Interest Rates and Its Application to Treasury Bond Options. *Financial Analysts Journal,* 46:33–39, 1990.

[4] F. Black and P. Karasinski. Bond and Option Pricing when Short Rates are Lognormal. *Financial Analysts Journal,* 47:52–59, 1991.

[5] F. Black and M. Scholes. The Pricing of Options and Corporate Liabilities. *Journal of Political Economy,* 81:637–659, 1973.

[6] W. E. Boyce and R. C. DiPrima. *Elementary Differential Equations.* Wiley, 9th edition, 2008.

[7] P. Boyle. A Lattice Framework for Option Pricing with Two State Variables. *Journal of Financial and Quantitative Analysis,* 23:1–12, 1988.

[8] P. Boyle, J. Evnine, and S. Gibbs. Numerical Evaluation of Multvariate Contingent Claims. *Review of Financial Studies,* 2:523–546, 1989.

[9] A. Brace, D. Gatarek, and M. Musiela. The Market Model of Interest Rate Dynamics. *Mathematical Finance,* 7:127–154, 1997.

[10] M. J. Brennan and E. S. Schwartz. A Continuous Time Approach to Pricing Bonds. *Journal of Banking and Finance,* 3:133–155, 1979.

[11] K. L. Chung and R. J. Williams. *Introduction to Stochastic Integration.* Birkhauser, Boston, 2nd edition, 1990.

[12] J. C. Cox, J. Ingersoll, and S. A. Ross. A Theory of the Term Structure of Interest Rates. *Econometrica,* 53:385–467, 1985.

[13] J. C. Cox and S. A. Ross. The Valuation of Options for Alternative Stochastic Processes. *Journal of Financial Economics,* 3:145–166, 1976.

[14] J. C. Cox, S. A. Ross, and M. Rubinstein. Option Pricing: A Simplified Approach. *Journal of Financial Economics,* 7:229–263, 1979.

[15] M. H. DeGroot and M. J. Schervish. *Probability and Statistics.* Pearson, 4th edition, 2011.

[16] D. Duffie. *Dynamic Asset Pricing Theory.* Princeton University Press, 3rd edition, 2001.

[17] D. Duffie and K. J. Singleton. *Credit Risk: Pricing, Measurement, and Management.* Princeton University Press, 2003.

[18] F. J. Fabozzi. *Fixed Income Mathematics: Analytical and Statistical Techniques.* McGraw-Hill, 4th edition, 2006.

[19] W. Feller. *An Introduction to Probability Theory and Its Applications, Volume I.* John Wiley, 1968.

[20] J. P. Fouque, G. Papanicolaou, and K. R. Sircar. *Derivatives in Financial Markets with Stochastic Volatility.* Cambridge University Press, 2000.

[21] D. T. Gillespie. *Markov Processes: An Introduction for Physical Scientists.* Academic Press, 1992.

[22] P. Glasserman. *Monte Carlo Methods in Financial Engineering.* Springer, 2003.

[23] M. J. Harrison and D. M. Kreps. Martingales and Arbitrage in Multiperiod Securities Markets. *Journal of Economic Theory*, 20:381–408, 1979.

[24] M. J. Harrison and S. R. Pliska. Martingales and Stochastic Integrals in the Theory of Continuous Trading. *Stochastic Processes and their Applications*, 11:215–260, 1981.

[25] D. Heath, R. Jarrow, and A. Morton. Bond Pricing and the Term Structure of Interest Rates: A New Methodology. *Econometrica*, 60:77–105, 1992.

[26] S. L. Heston. A Closed-Form Solution for Options with Stochastic Volatility with Applications to Bond and Currency Options. *The Review of Financial Studies*, 6(2):327–343, 1993.

[27] T. S. Y. Ho and S. B. Lee. Term Structure Movements and Pricing Interest Rate Contingent Claims. *Journal of Finance*, 41:1011–1029, 1986.

[28] J. Hull. *Options, Futures, and Other Derivatives.* Pearson, 9th edition, 2014.

[29] J. Hull and A. White. The Pricing of Options on Assets with Stochastic Volatilities. *Journal of Finance*, 42:281–300, 1987.

[30] J. Hull and A. White. Pricing Interest Rate Derivative Securities. *Review of Financial Studies*, 3:573–592, 1990.

[31] K. Ito. On a Formula Concerning Stochastic Differentials. *Nagoya Mathematics Journal*, 3:55–65, 1951.

[32] P. Jackel. *Monte Carlo Methods in Finance*. Wiley, 2002.

[33] R. A. Jarrow and A. Chatterjea. *An Introduction to Derivative Securities, Financial Markets, and Risk Management*. W. W. Norton & Company, 2013.

[34] I. Karatzas and S. E. Shreve. *Brownian Motion and Stochastic Calculus*. Springer, 2nd edition, 1991.

[35] F. A. Longstaff and E. S. Schwartz. Interest Rate Volatility and the Term Structure. *Journal of Finance*, 47:1259–1282, 1992.

[36] D. G. Luenberger. *Investment Science*. Oxford Press, 2nd edition, 2013.

[37] W. Margrabe. The Value of an Option to Exchange One Asset for Another. *Journal of Finance*, 33(1):177–186, 1978.

[38] R. L. McDonald. *Derivatives Markets*. Pearson, 3rd edition, 2013.

[39] R. C. Merton. The Theory of Rational Option Pricing. *Journal of Economics and Management Science*, 4(1):141–183, 1973.

[40] R. C. Merton. Option Pricing When the Underlying Stock Returns are Discontinuous. *Journal of Financial Economics*, 5:125–144, 1976.

[41] K. Miltersen, K. Sandmann, and D. Sondermann. Closed Form Solutions for Term Structure Derivatives with Log-Normal Interest Rates. *Journal of Finance*, 52(1):409–430, 1997.

[42] B. Oksendal. *Stochastic Differential Equations: An Introduction with Applications*. Springer, 5th edition, 1998.

[43] R. Rendleman and B. Bartter. Two State Option Pricing. *Journal of Finance*, 34:1093–1110, 1979.

[44] L. C. G. Rogers and D. Williams. *Diffusions, Markov Processes and Martingales: Volume 1, Foundations*. Cambridge University Press, 2000.

[45] L. C. G. Rogers and D. Williams. *Diffusions, Markov Processes and Martingales: Volume 2, Ito Calculus*. Cambridge University Press, 2000.

[46] S. A. Ross. The Arbitrage Theory of Capital Asset Pricing. *Journal of Economic Theory*, 59:341–360, 1976.

[47] P. J. Schonbucher. *Credit Derivatives Pricing Models: Models, Pricing and Implementation*. Wiley, 2003.

[48] S. Stein and A. Barcellos. *Calculus and Analytic Geometry*. McGraw-Hill, 5th edition, 1992.

[49] G. Strang. *Linear Algebra and Its Applications*. Brooks Cole, 4th edition, 2006.

[50] G. E. Uhlenbeck and L. S. Ornstein. On the Theory of Brownian Motion. *Physical Review*, 36:823–841, 1930.

[51] O. Vasicek. An Equilibrium Characterization of the Term Structure. *Journal of Financial Economics*, 5:177–188, 1977.

[52] R. G. Vickson. An Intuitive Outline of Stochastic Differential Equations and Stochastic Optimal Control. In *Stochastic Models in Finance*. World Scientific, 2nd edition, 2006.

[53] P. Wilmott. *Paul Wilmott on Quantitative Finance*. Wiley, 2nd edition, 2006.

Index

Printed and bound by CPI Group (UK) Ltd, Croydon, CR0 4YY

23/10/2024

01778004-0001